Proceedings of the Royal Institution of Great Britain

Proceedings of the Royal Institution of Great Britain

Volume 71

Edited by
RICHARD CATLOW
and
SUSAN GREENFIELD

UNIVERSITY PRESS

OXFORD
UNIVERSITY PRESS

Great Clarendon Street, Oxford OX2 6DP

Oxford University Press is a department of the University of Oxford.
It furthers the University's objective of excellence in research, scholarship,
and education by publishing worldwide in

Oxford New York

Athens Auckland Bangkok Bogotá Bombay Buenos Aires
Cape Town Dar es Salaam Delhi Florence Hong Kong Istanbul
Karachi Kalkota Kuala Lumpur Madrid Melbourne Mexico City Mumbai
Nairobi Paris São Paulo Shanghai Singapore Taipei Tokyo Toronto Warsaw

and associated companies in Berlin Ibadan

Oxford is a registered trade mark of Oxford University Press
in the UK and in certain other countries

Published in the United States
by Oxford University Press Inc., New York

First published 2001

A catalogue record for this title is available from the British Library

Library of Congress Cataloging in Publication Data
(Data available)

ISBN 0 19 850992 8

10 9 8 7 6 5 4 3 2 1

Typeset by EXPO Holdings, Malaysia

Printed in
on acid-free paper by
Biddles Ltd.,
Guildford & King's Lynn

CONTENTS

CONTRIBUTORS

Clifford Friend
Cranfield University
RMCS Shrivenham

Richard Gregory
Emeritus Professor of Neuropsychology
University of Bristol

John M. Holloway
Department of Chemistry
University of Leicester

Kenneth Ives
Department of Civil and Environmental Engineering
University College London

Nancy J. Lane
Department of Zoology
University of Cambridge

Charles Taylor
Former Professor of Experimental Physics
The Royal Institution, London

John M. Taylor
Director General of Research Councils

H. J. V. Tyrell
Former Chairman of the Buildings Working Party
The Royal Institution, London

Arnold Wolfendale
Former Professor of Experimental Physics
The Royal Institution, London

Robert M. Worcester
Chairman MORI/Visiting Professor of Government
London School of Economics and Political Sciences

Cosmic rays

ARNOLD WOLFENDALE

Introduction

The end of the nineteenth century saw many studies of the properties of ionizing radiations, and the names of the most important practitioners such as Becquerel, the Curies and Röntgen are now part of the language of science.

One problem, that survived the dawn of the new century, was the reason for the decay of charge on the electroscopes even when no obvious radiation was present. One idea was that there was some form of super-gamma-radiation coming from unknown sources in the earth, and a number of experiments were mounted on high towers to check the hypothesis. No clue for the origin of this 'new' radiation appeared, however, and it was natural that balloon-borne equipment should be pressed into service. A number of comparatively low-level flights followed, but it was not until an ardent and skilful balloonist, Viktor Hess, achieved sufficient height (5.3 km) and with a superior electroscope, that success was achieved. A series of three flights, in 1912,[1] gave consistent results—for the strength of the radiation: the intensity fell slightly and was then roughly constant to a height of 2 km but it more than doubled by a height of 5 km. Hess then made his famous claim:

> *The results of the present observations seem to be most readily explained by the assumption that a radiation of very high penetrating power enters our atmosphere from above, and still produces in the lowest layers a part of the ionization observed in closed vessels.*

In the next two years he went even higher and reached 9 km, where the increase was a factor of about 7. Kilhorster[2] in 1913 and 1914 went higher still, with similar results (Fig. 1). Nowadays his conclusion is an obvious one, but this was not so then when not only were detectors in a far-from-perfect state but atmospheric phenomena could well have been

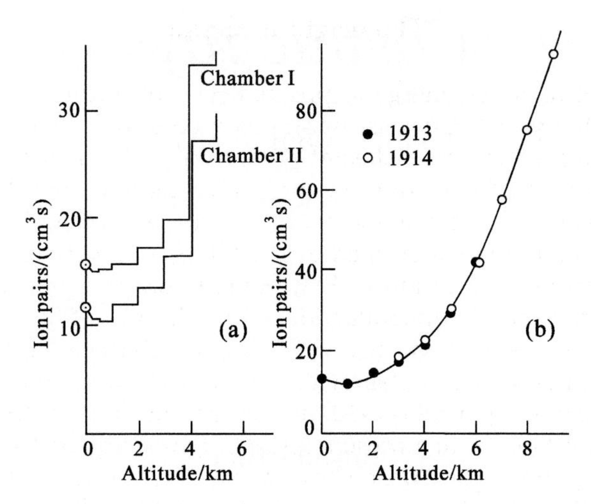

Fig. 1 Variation of ionization with altitude (a) Hess (1912); (b) Kolhörster (1913, 1914).[1,2]

responsible. Indeed, had the presence of the ionosphere been known at that time, it is very likely that Hess's conclusions would have been very different.

It seems that it was Millikan, of 'charge-on-the-electron' fame, who coined the term 'cosmic radiation' to describe the incoming entities. In fact, it was an unfortunate title because the vast majority of the incoming 'radiation' turns out to be nuclear particles (including electrons and the elusive neutrinos).

Later work has identified a great range of nuclear charges in the beam, at least at 'low' energies, where direct measurements can be made.

Measurements in the 1930s and 1940s—using nuclear emulsions and cloud chambers—led to the discovery of many important fundamental particles, amongst them the pion, muon, positron, and the 'strange particles'. With the advent of accelerators in the 1950s the field of elementary-particle physics has largely been the province of the accelerator physicist although there have been advances, from time to time, from studies using cosmic rays—at energy above those available at the big machines[3] and from measurements on neutrinos.

The subject has, nevertheless, advanced considerably in the area of astrophysics and, to a much lesser extent, in the field of geophysics. In astrophysics there are two aspects: the use of astronomical phenomena to shed light on the problem of cosmic ray origin and the reverse, the use of cosmic ray results to illuminate astronomical problems. Both will be considered here.

The origin problem

Most cosmic rays are energetic protons (we discount neutrinos, from here on, although neutrino astronomy is a coming science). Having an electrical charge they are deflected by the magnetic fields encountered between source and detection on earth. Moving out from the earth, the magnetic fields are those associated with the earth, the interplanetary medium, and the interstellar medium (ISM). If, as is very likely, the most energetic particles of all are of Extragalactic origin, then the list must include the field in the intergalactic medium (IGM). Of these fields, only the first mentioned is known with any precision. Digressing for a moment into the domain of 'energy', cosmic rays start, conventionally, at the rest mass of the particles ($\sim 10^9$ eV for protons) and continue up, with diminishing intensity, to a present record high energy of 3×10^{20} eV (the Fly's Eye experiment).[4] Those above about 10^{11} eV are not affected by the field in the interplanetary medium, but there is then a vast expanse of energy—to ~ 10^{18} eV—where the field in the ISM dominates, if the primaries are still mainly protons. The expanse extends to higher energies still, if the nuclei are more massive (e.g. times 26 for iron primaries). Figure 2 gives the cosmic-ray energy spectrum at earth for some of the major components.

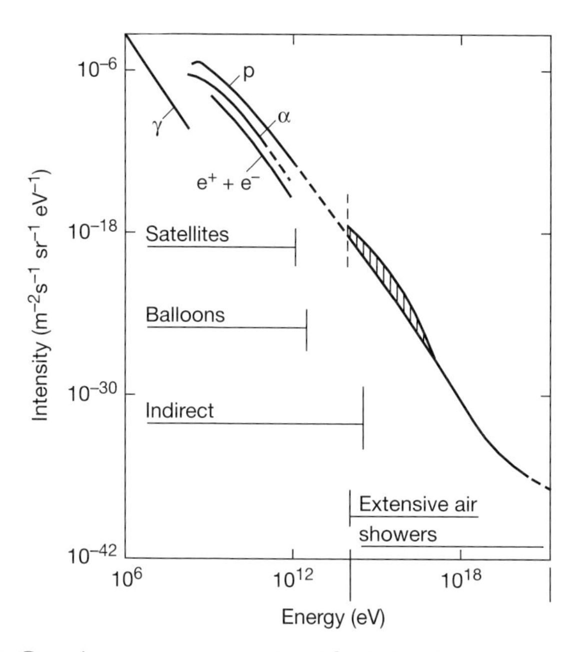

Fig. 2 Cosmic-ray energy spectra. The intensity is uncertain in the shaded area—although Erlykin and I think we know what it is! The methods of study are indicated.

Only at the very lowest energies (~ < 10^{10} eV) do cosmic rays occasionally arrive from the sun, largely via the giant solar flares. Thus, one is immediately into the problem of the magnetic field in the ISM. Alas, its topography is not known in any detail at all so that 'ray-tracing', from detector back to source, is impossible (the situation for that tiny fraction of the cosmic ray flux above 10^{18} eV will be considered later). Other arguments are necessary.

Before continuing, we can identify two (extreme) possibilities for cosmic ray origin: that they are Galactic or that they are Extragalactic.

Energy densities

It is instructive in all astrophysical systems, and, indeed, elsewhere, to consider the energy density of the phenomenon under study in comparison with those of possibly related phenomena. Table 1 gives such a comparison. Considering the Galactic energy densities, it looks as though there might be a case for concluding that 'cosmic rays are Galactic', although it is rather off-putting that apparently disparate phenomena, such as stellar radiation and gas motion, have the same energy density as that in cosmic rays. For the Galactic magnetic field, however, a plausible connection with cosmic rays can be made, although this might be more to do with trapping than with origin itself. The argument here is that the cosmic ray pressure builds up until it is equal to that in the magnetic field and then the field 'expands' to allow some particles to leak out into extragalactic space.

This argument with respect to a Galactic origin breaks down when it is realized that the energy density of the cosmic microwave background (CMB)—which is certainly of Extragalactic origin—is the same as that in

Table 1. Energy densities in the local Galaxy

Phenomenon	eV cm^{-3}
Cosmic rays $\left(\frac{4\pi}{c}\int E\, I(E) dE\right)$	~ 0.5
Magnetic field $(B^2 / 8\pi)$	~ 0.5
Starlight $\frac{4\pi}{c}\int h\nu\, I(h\nu)\, d\, h\nu$	~ 0.5
Cloud motion $\left\langle \Sigma \frac{1}{2} M v^2 \right\rangle / V$	~ 0.5

cosmic rays, to within a factor 2 (it is 0.24 eV cm^{-3}). Bearing in mind that most of the cosmic rays (and the associated energy density) are below about 10^{10} eV—and well below the energy where interactions between cosmic rays (protons) and the CMB start to become important (~2.10^{18} eV)—an Extragalactic origin for cosmic ray protons cannot be ruled out. It is true (see, for example, reference 5) that although the energy content in cosmic rays in the Universe as a whole would be high, it would not be impossibly so; it is also true that there are potential sources in Extragalactic space.[5]

Another approach must be made.

Gamma-ray astronomy

Are cosmic rays Galactic or Extragalactic?

Although the bulk of cosmic rays are particles, about 10^{-6} of the flux is composed of gamma rays. In so far as many of the gamma rays are produced by the interactions of cosmic rays with the ISM, and they travel in straight lines, they provide proxy indicators of cosmic ray intensities in other parts of the Galaxy and beyond. Several approaches have been made, and these lead to the conclusion that particles of Galactic origin predominate.

The first, historically,[6] (see also a summary[7]) was the demonstration of a 'Galactic gradient', i.e. a fall-off of cosmic ray intensity with distance from the centre of the Galaxy. Although subject to considerable argument, initially, it is now accepted that there is such a gradient, indicating that the sources of most cosmic rays are in our own Galaxy and there are more and more as one approaches the Galactic Centre. An interesting—and important—by-product of the early work was the demonstration that the mass of molecular hydrogen in the ISM had been over-estimated[8] by those whose measured carbon monoxide line intensities had been used, via an empirical constant, to estimate the column density of H_2. This 'discovery' is a good example of the use of cosmic rays to help the astronomers.

Confirmation of the fact that most of the cosmic rays in the Galaxy are actually produced here has come from the detection of gamma rays from the Magellanic Clouds (MC). The argument is a simple one. If cosmic rays were Extragalactic, specifically if they were Universal—i.e. filled the Universe at roughly the same intensity—then the intensity within the MC would be the same as that locally. Since the mass of target gas in the MC is known reasonably well, the expected flux of gamma rays can be calculated. Comparison with observation then enables the question of

Universality, or otherwise, to be answered. Whilst it is true that, as usual, there are subtleties, in this case largely to do with the contribution to the gamma ray flux from unresolved discrete sources, the conclusion is quite robust. It is (Fig. 3) that the observed fluxes from both Magellanic Clouds, Large and Small, are much smaller than would be the case for a cosmic ray intensity there being the same as that locally.[9,10] It is concluded that less than 10% of cosmic rays at earth come from Extragalactic sources. In fact, the fraction is very probably considerably less than 10%. It must remember, however, that the figure given above relates to the total intensity and most of this is in low-energy particles (Fig. 2). At high energies, say above 10^{18} eV, it could well be that *most* of the particles come from beyond the Galaxy. This topic will be taken up in more detail, later.

The question now, is, just what are the Galactic objects, or systems, responsible for the 'low'-energy particles? There are a number of possibilities, but supernova remnants (SNR) are the front-runners. The idea, which has been worked out in many publications,[11,12] is that the SNR shocks, passing through the ISM, accelerate the 'seed' cosmic rays to high energies. In a standard calculation,[12] the energy spectrum generated

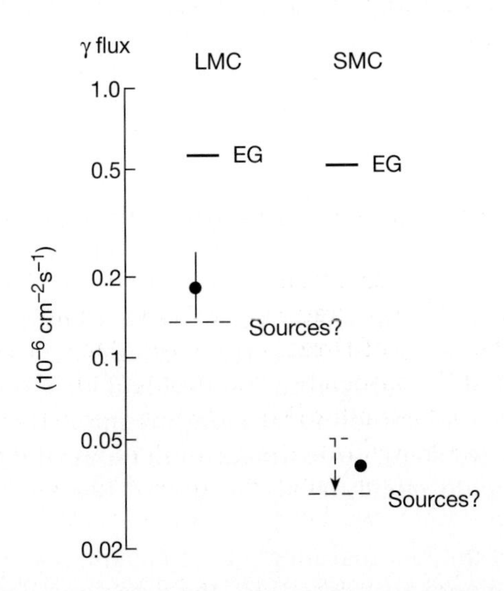

Fig. 3 Gamma-ray fluxes (of energy above 100 MeV) from the GRO satellite in comparison with what would be expected if the cosmic ray intensity were the same there as locally (denoted EG).[9,10] The observed fluxes are much smaller than expected, indicating that most of the local cosmic rays come from our own Galaxy. (The estimated 'source' contributions are indicated.)

is roughly of the form $N(E) = A\,E^{-2}$ where E is the (relativistic) energy, and the maximum energy is given by:

$$E_{\max}(\mathrm{GeV}) = 4\times10^{5}Z\left(\frac{E_{\mathrm{SN}}}{10^{51}\,\mathrm{erg}}\right)^{1/2}\left(\frac{M}{10M_{\mathrm{O}}}\right)^{-1/6}\left(\frac{N_{\mathrm{O}}}{3\times10^{-3}\,\mathrm{cm}^{-3}}\right)^{-1/3}\left(\frac{B_{\mathrm{O}}}{3\mu G}\right) \quad (1)$$

where E_{SN} is the SN energy, M the released mass, N_{O} the gas density, and B_{O} the magnetic field.

Since the most favourable value for all the parameters in brackets is ~1 (and this for the 'hot' ISM), it is seen that the maximum for iron nuclei, the most likely 'massive' contributor to cosmic rays (for good astrophysical reasons), is a little over 10^{16} eV. A. D. Erlykin and I claim to have discovered a feature in the cosmic ray spectrum which can be attributed to iron nuclei of this order of energy (and other nuclei, too), and the very recent results from our work will be described shortly.

Irrespective of the Erlykin and Wolfendale (EW) work, most cosmic ray workers would agree with the argument that SNR are powerful contributors to the general cosmic ray flux—at least to 10^{15} eV, or so. At higher energies there is, alarmingly, a dearth of good origin theories.

Before delving into the EW results, an analysis will be described of another, powerful, application of the cosmic gamma-ray technique—the search for anti-matter.

Gamma rays and the anti-matter problem

Physics has many 'conservation laws'—such as the conservation of energy, momentum, charge, and so on. With the advent of anti-matter, predicted from the work of Dirac, and particularly apparent in the case of electrons (e^-) and the associated positrons (e^+), it was natural to expect that matter (M) and anti-matter ($\overline{\mathrm{M}}$) were conserved in the sense that for every unit of matter created there was an equal amount of anti-matter. If, as would be expected, equal amounts of M and $\overline{\mathrm{M}}$ were created in the very early Universe then we have the big question: 'Where has all the anti-matter gone?'.

We know from rather direct arguments that there can be no question of there being matter–anti-matter symmetry in the solar system (Neil Armstrong's words when he stepped on the moon were not his *last* words: 'One small step for man, 10^{30} pi-mesons for mankind'!). Many studies have shown that the gamma rays resulting from $\mathrm{M}\overline{\mathrm{M}}$ annihilation would have given too high a flux to be compatible with observation if there were symmetry on any scale out to that of galaxy clusters.

However, there is (or, rather, was) the possibility that on the very largest scales in the Universe—that of superclusters of galaxies—there was such symmetry. The idea is that in the very early Universe, perhaps, cells of M and $\bar{M}$ were created which rapidly separated. We, ourselves—by definition—reside in a matter-supercluster.

Dr Dudarewicz and I[13] looked at this topic in some detail, using the latest cosmic gamma ray and astronomical data, and we concluded that there would be an overlap of the hot gas, and $\overline{\text{gas}}$ escaping from the superclusters (S) and anti-superclusters ($\bar{S}$) such as to still give too many gamma rays (see Fig. 4). Another interesting feature was the fact that occasionally an individual galaxy (or $\overline{\text{galaxy}}$) would venture into the opposite regime. Again, too many gamma rays would be produced. Figure 4 shows the results.

Thus, there is no $M\bar{M}$ symmetry on any scale. Just how the $\bar{M}$ disappeared soon after the Big Bang is still not completely clear.

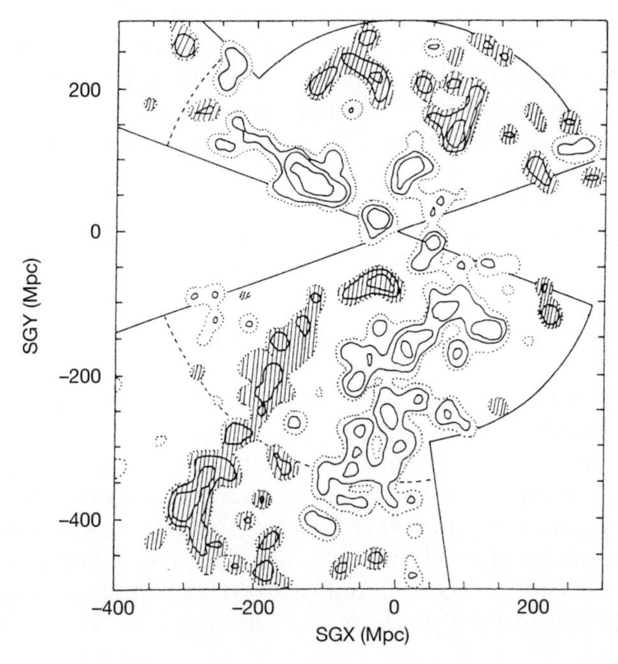

Fig. 4 Map of the nearby Universe with marked typical divisions into superclusters (shaded) and anti-superclusters.[13] (umshaded) Our analysis shows that such divisions cannot exist—on any scale—as there would be too big a flux of extragalactic gamma rays from the gas and anti-gas annihilating. Thus the contemporary Universe must be overwhelmingly of matter, and not anti-matter. We are at the centre (0, 0) on the edge of the local (Virgo) supercluster.

Cosmic rays and life on earth

The 'normal situation'

As a source of the radiation 'background' to which humans are subjected, cosmic rays are not negligible, about 30% of the natural background being due to this cause (near ground level and in 'average' locations). The rest is due to the radioactive emanations from naturally occurring materials.

However, cosmic rays *may* also have relevance to another phenomenon which affects the human condition: the climate. The situation relates to the apparent 'effect of sunspots on the weather'. In fact, there can be no direct effect because the energy changes in solar irradiation associated with the 11-year solar cycle are quite negligible. That is not to say that individual wavelength regions do not vary much—the reverse is true: the UV flux, for example, changes considerably (by about 30% over the 11-year cycle). However, the energy in this component is absorbed in the upper levels of the atmosphere, and there has been the traditional view that there is very little interchange between the upper levels and the lower region (troposphere) where 'the weather' resides. Help may now be at hand with the claim (which looks quite well-founded) that the link between the solar variability and the troposphere may be via cosmic rays. The argument runs as follows. Cosmic rays produce ionization of the air as they pass through and ionization is relevant to cloud-forming ability. The cosmic rays of concern are mainly of Galactic origin (from supernovae—see later) and their intensity is modulated considerably by

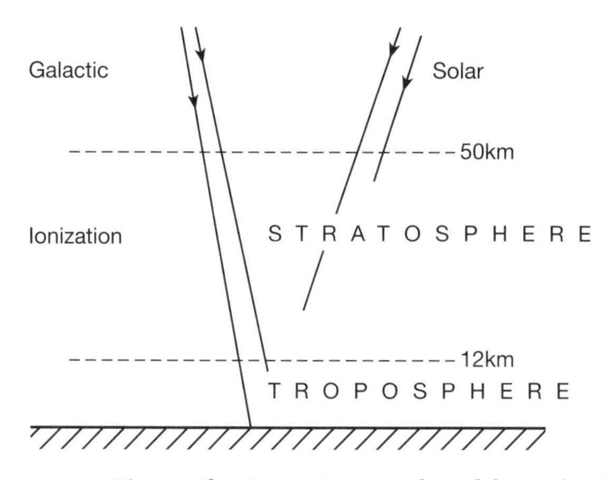

Fig. 5 The ionization produced by Galactic cosmic rays, and solar flare particles, which might have an effect on the climate.

the solar wind over the 11-year cycle. The evidence so far is a correlation[14] between the cloud cover over the oceans and the cosmic ray intensity. There is also a claim that the ozone density in polar regions is affected by low-energy solar protons, the intensity of which varies, again, during the solar cycle.[15] It is true that there is argument as to whether the ionization effect is large enough to cause the observed effects—cloud cover and ozone depletion—but, at least, it gives a starting point for more detailed analysis.

It is also true that the effect on long-term climate, as distinct from local weather, remains to be established; however, the stakes are so high in this field that serious study is essential.

Rare solar phenomena

It is not outside the realms of possibility that very rare solar flares, and associated cosmic ray particles, may have a significant effect on 'life on earth'. Figure 6 shows the frequency distribution for the energy deliv-

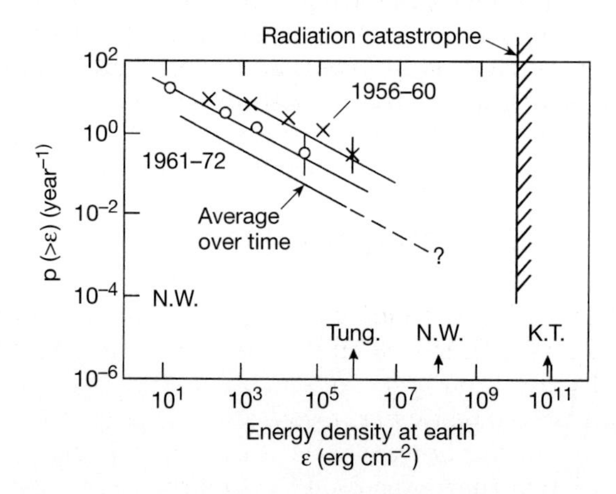

Fig. 6 Cosmic rays associated with solar flares.[16] $p\,(>\varepsilon)$ denotes the frequency, per year, of energetic solar flares which deposit more than energy ε at the earth. The points come from observations on solar flares for the periods indicated. 'Average over time' relates to the average over many decades and, specifically, for the energy deposited on the earth. Tung., N. W. and K. T. denote energy densities associated with the Tunguska bolide, the 'Nuclear Winter' (an East–West exchange of 10 000 megatons each), and the (possible) bolide which arrived 65 Ma ago at the K–T, geological boundary.

ered to earth by cosmic ray flares.[16] Measurements have been made for only 50 years or so and extrapolations are clearly dangerous, but it would not be at all surprising if every few hundred thousand years, or so, there were quite damaging flare effects. The fact that we are here at all means that really catastrophic flares have not occurred but that does not mean that serious effects have been absent nor does it mean that serious effects will not occur in the future. Although some work has been done on this topic it would repay further study.

Fine structure in the cosmic ray spectrum

The problem

So far, the evidence favouring cosmic ray production in supernova remnants (SNR) has been circumstantial. What, then, would be a distinctive signature? At very low energies ($E \sim< 10^{10}$ eV) there is the gamma ray work which shows a small excess of cosmic ray intensity within the Loop I SNR,[17] although the reason for the excess is not completely secure. If it could be shown that there is a signal consistent with the theoretical predictions (see earlier) of a sharp cut-off at an energy in the region of 4×10^{15} Z eV then the case for the SNR region would be strengthened considerably.

The EW analysis

Tolya Erlykin and I have made such a claim[18] from a detailed study of extensive air showers (EAS). Such showers were discovered in 1937 by Auger and Maze and their associates[19] and are due to the impact on the atmosphere of very energetic particles. Figure 7 shows the situation. It is the generation of such showers, with their wide lateral spread (many hundreds of metres) that enables the extraordinarily rare high-energy particles to be detected at all. Paradoxically, to be deep in the atmosphere is better for detection than to be at great heights!

One of the main quantities measured in EAS is the so-called size spectrum, $F(N)$, i.e. the frequency with which a shower of size N particles is recorded (N is determined by allowing for those particles which fall between and outside the particle detectors which are spread over the ground).

Figure 7 shows the 'size spectrum' in diagrammatic form. Some forty years ago[20] a 'knee' was discovered—a feature which has persisted to

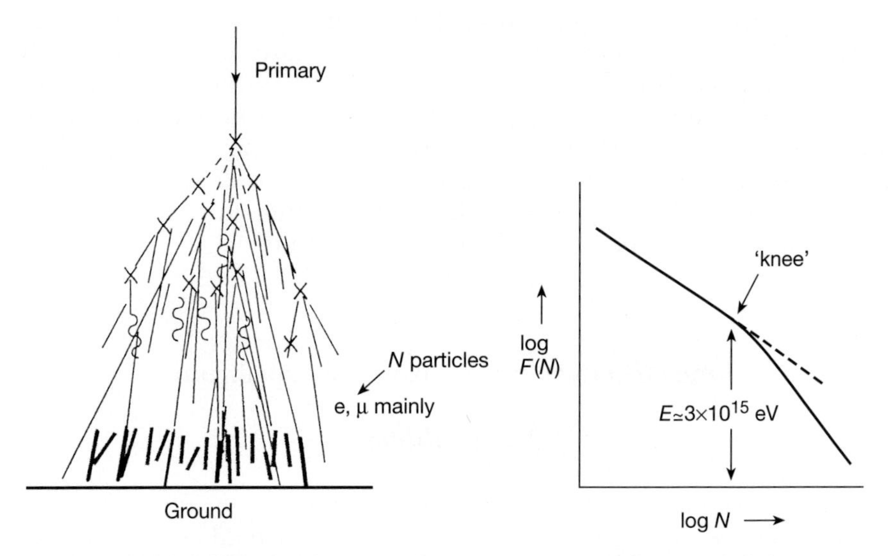

Fig. 7 Schematic diagram of an 'extensive air shower' and the corresponding 'size spectrum'. N denotes the total number of particles in the shower at ground level. The knee, discovered by the Moscow State University (MSU) group in 1958,[19] is indicated.

this day. It has been found time and time again by detection arrays all over the world. Clearly, the knee is trying to tell us something about where the particles come from and how they have been accelerated. It is this feature that Erlykin and I claim to have explained.

A sample of the world's data on $F(N)$ is given in Fig. 8; there is roughly the same amount of information, again, from more recent experiments. All of it, to our way of thinking, shows the same result: a small 'peak' in the size spectrum at 0.6 in log N beyond the knee. Our results stand, or fall, on this identification.

It will be noted from the figure that occasionally there are small displacements from 0.6; we claim to understand the reason for most of them. Essentially, in any one experiment, the peak is not statistically significant but when all are added together it becomes so.

Fig. 8 Size spectra from many experiments, standardized to the same knee position (size = $N_e$knee).[17] We (Erlykin and Wolfendale) claim to have identified a small 'peak' at an abscissa of +0.6. We associate the knee with oxygen nuclei and the second 'peak' with iron nuclei.

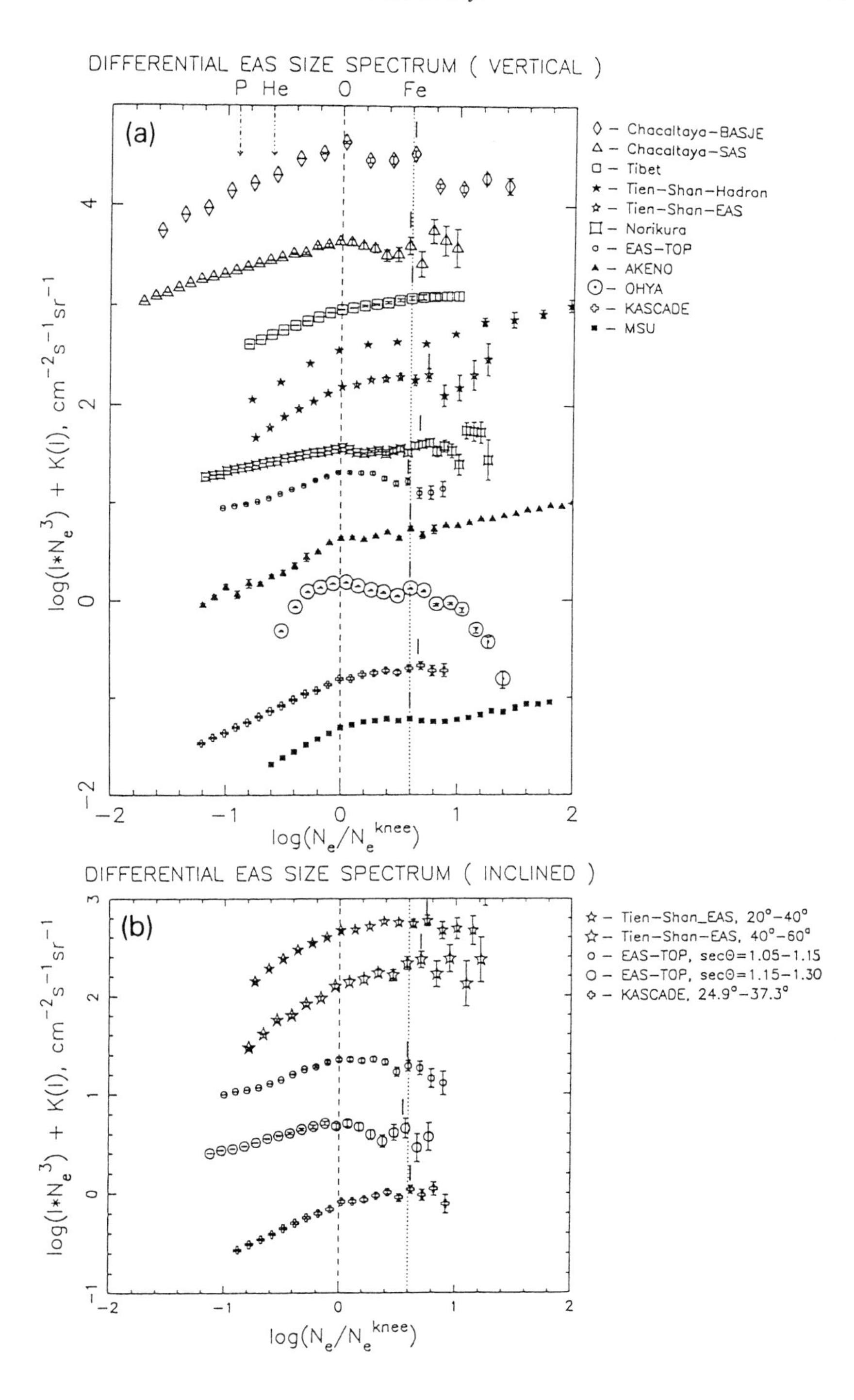
DIFFERENTIAL EAS SIZE SPECTRUM (VERTICAL)
P He O Fe
(a)
◊ – Chacaltaya–BASJE
△ – Chacaltaya–SAS
□ – Tibet
★ – Tien–Shan–Hadron
☆ – Tien–Shan–EAS
Norikura
EAS–TOP
AKENO
OHYA
KASCADE
MSU
log(I*Ne^3) + K(I), cm^-2 s^-1 sr^-1
log(Ne/Ne^knee)
DIFFERENTIAL EAS SIZE SPECTRUM (INCLINED)
(b)
Tien–Shan_EAS, 20°–40°
Tien–Shan–EAS, 40°–60°
EAS–TOP, secΘ=1.05–1.15
EAS–TOP, secΘ=1.15–1.30
KASCADE, 24.9°–37.3°

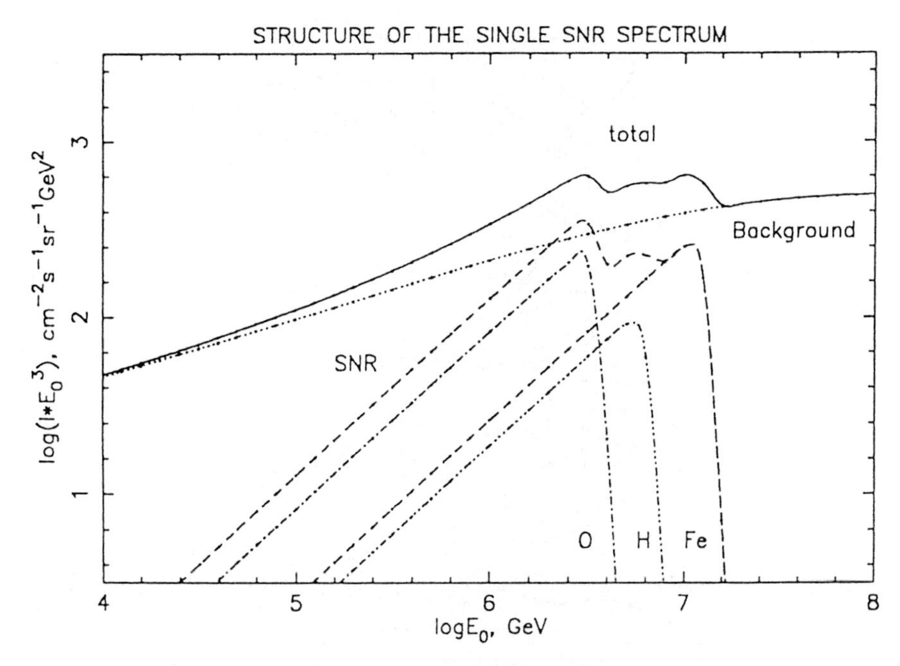

Fig. 9 The Erlykin and Wolfendale energy spectra from the hypothesized local supernova remnant (SNR).[17] H represents 'heavy nuclei' in the range Ne–S.

The EW interpretation

Figure 9 shows the interpretation, in terms of energies, that the energy spectrum comprises two components, one from a 'background'—due to very many, widely distributed, sources in the Galaxy—and the other due to a single, local, recent supernova. This latter should produce the groups of nuclei shown and their spectra should cut off just where we need them *if* the SN exploded in the local 'hot' ISM, for which the equation (1) is valid.

The cosmic ray jigsaw

In a number of recent papers we have examined the extent to which a wide variety of cosmic ray phenomena fit together (Fig. 10). We claim that each and every phenomenon is at least not inconsistent with our claim; indeed, some back it very strongly.

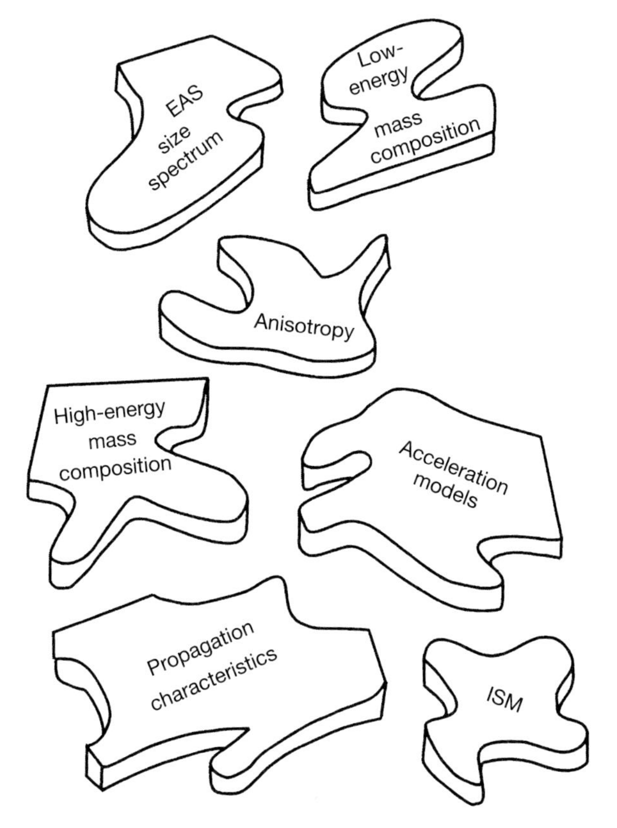

Fig. 10 The cosmic ray jigsaw. The Erlykin and Wofendale model seems to explain them all and enables them to fit together.

Where is the actual supernova?

The final question—and it is one that we have not yet solved—is the actual location of the claimed supernova. Although there are several possibilities ('Geminga', Vela, and perhaps an undetected one in the Loop I region) we have yet to make a definitive pronouncement. Time will tell.

The future

Not only do we need more data in the region of the knee in the spectrum but the source of the 'background' spectrum still needs explaining. Finally, there is the region above 10^{18} eV. It is likely that many of the particles below 10^{19} eV are iron nuclei produced in our own Galaxy but, at higher energies still, it is hard to escape the conclusion that the particles are Extragalactic in origin. Here, as already mentioned, the uncertain magnetic field in the IGM causes problems. The field *should* be small enough to allow 'local' (say within 100 million light years) sources to be identified but none have been, as yet, with any degree of certainty. The hunt continues.

References

1. Hess, V.F., *Phys. Z.,* 1912, **13**: 1084.
2. Kollhörster, W, (1913, 1914) quoted in *Cosmic Rays* by J.G. Wilson, The Wykeham Science Series, Taylor & Francis, 1976, London.
3. Wdowczyk, J. and Wolfendale, A.W., *J. Phys. G. Nuc. Phys.,* 1987, **13**: 411.
4. Bird, D.J. *et al., Astrophys. J.,* 1995, **441**: 144.
5. Wolfendale, A.W., *Q. Jl. R. Astr. Soc.,* 1983, **24**: 122.
6. Dodds, D., Strong, A.W., and Wolfendale, A.W., *Mon. Not. R. Astr. Soc.,* 1975, **171**: 569.
7. Ramana Murthy, P.V. and Wolfendale, A.W., *Gamma Ray Astronomy,* Cambs. Astrophys. Ser., Cambridge University Press, 1986, 1993.
8. Bhat, C.L. *et al., RAL Workshop on Astron. and Astrophy.*, 1984, RAL-84-101 (Gondhalekar, P.M., ed.) p. 39.
9. Sreekumar, P. *et al., Astrophys. J.*, 1992, **400**: L67.
10. Chi. X. and Wolfendale, A.W., *J. Phys. G.*, 1993, **19**: 795.
11. Axford, W.I., *Proc. 17th Int. Cosmic Ray Conf. (Paris)*, 1981, **12** (Saclay: CEN) p. 155.
12. Berezhko, E.G., *Astropart. Phys.*, 1996, **5**: 367.
13. Dudarewicz, A. and Wolfendale, A.W., *Mon. Not. R. Astr. Soc.*, 1994, **268**: 609.
14. Svensmark, H. and Friis-Christensen, *Journ. Atmos. and Terr. Phys.*, 1997, **59**: 1225.
15. Shumilov, O.I. *et al., Journ. Atmos. and Terr. Phys.*, 1995, **57**: 665.
16. Wdowczyk, J. and Wolfendale, A. W., *Nature*, 1977, **268**: 510.
17. Bhat, C.L. *et al., Nature*, 1985, **314**: 515.
18. Erlykin, A.D. and Wolfendale, A.W., *Astropart. Phys.*, 1998, **8**: 265.
19. Auger, P., Maze, R., and Robley, C.R., *Compt. Rend.*, 1938, **208**: 1641.
20. Kulikov, G.V. and Khristiansen, G.B., *JETP*, 1958, **35**: 635.

A. WOLFENDALE

Born in 1927 in Rugby, he studied physics in Blackett's famous laboratory at Manchester University. After graduation in 1948, he started his research in cosmic ray physics and has been at it ever since, although his interests have broadened into astrophysics in general. Moving from a Lectureship in Manchester to Durham University in 1956 he was Head of Department for a number of spells and was responsible for initiating the astronomy and astrophysics research effort there. He was President of

The Royal Astronomical Society 1981–3, the Durham University Society of Fellows, 1988–94, the Institute of Physics, 1994–6, and the European Physical Society 1999–2001. He was the 14th Astronomer Royal 1991–5. Particularly pleasing to him was his appointment as Professor of Experimental Physics at the Royal Institution, in 1996. He was elected Fellow of the Royal Society in 1977. He holds Honorary degrees from a number of Universities, both in the UK and abroad and he is a Foreign Fellow of Academics in India and South Africa. He was knighted in 1995.

Materials for 'life'

CLIFFORD FRIEND

A few words of introduction

The civilization of humankind has been intimately linked with the use of materials. From the days of 'stone' to the refining of today's alloys and silicon chips, our lives have become enriched by technologies which depend critically on the production, design, and use of structural and 'functional' materials. However, surprisingly, for most of these thousands of years humankind has exploited engineered materials without really knowing much about them. Instead, materials developments have depended on a craft tradition followed perhaps by what can only be described as empiricism. I hope you will detect in this paper my enthusiasm for materials engineering. This comes partly from the fact that much of our core science is so relatively new, many of the idols and 'gods' of our profession still being alive or having been with us until relatively recently. It is these people who have told us about the fundamental science controlling the performance and application of materials. As well as our recent science there is one other important feature of materials engineering. We now understand our science more and more, but what really obsesses us is the desire for applications. In other words we love to see products in the world around us where our science is clearly at work. To some extent the latter is what this paper is about. However, I first want to say something more about the mind-set of materials engineers.

As with most engineers, materials professionals have been obsessed with the big and grandiose application of our technologies. Let's build an aircraft out of this wonderful new structural material! Or, what about bridges? This is the sort of application world in which most materials engineers live and work. What I want to do in this paper is to change the viewpoint slightly. To look at a different world of materials applications. To date we seem to have developed our wonderful science with a focus primarily on large engineering structures. However, it is equally applicable to areas which have been described as 'less traditional'. It is these which I will concentrate on here.

My title—'Materials for "life"'—is a sad attempt by a materials engineer to be clever, so please bear with me. You will soon encounter a number of puns based on this title. However, I hope you will see that they do make important points about the impact materials technologies have on almost all aspects of our everyday life. So with our introduction complete we should embark on our journey through materials science and its effect on 'non-traditional' areas close to our everyday life. As a final introductory remark I hope you, the reader, will bear with me if I am a little selfish. We could discuss many people's work but I will concentrate here on my own team's research. This is quite deliberate since this work is very personal to us and I hope it will communicate to you some of our enthusiasm for it and the potential of its application.

'Life' saving

The first area I wish to explore with you is health-care. We hear about drug research and all the science connected with it, but what new developments can be brought to the field of health-care from materials science—things that can make a very major impact on the health of our community? Let's first consider breast cancer. You may not be aware that one in twelve women in the UK will experience breast cancer during her life—the highest incidence in the world. Can materials science assist us in the fight against such cancers? What can materials science do in a field that looks so clearly medical? Let's look first at the diagnosis of breast cancer. Figure 1 shows a mammogram: an X-ray photograph of the breast.

Many female readers may have experienced mammography, which is the way breast cancer is currently diagnosed during screening programmes. Radiologists view such mammograms and look for the presence of features which can be interpreted as premalignancy or malignancy and then carry out biopsies for final pathological diagnosis. As the incidence of this cancer is so high there are massive numbers of women who will experience this condition and it is critical that we diagnose as early as possible so that their prognosis is good. So what can materials science offer us in this field?

We are using a new approach to improve the quality of mammograms which we hope will allow earlier detection of premalignant features within the breast. We are doing this using a special form of X-rays known as 'synchrotron radiation'. So what is new about this? These X-rays are intense and have a very narrow range of wavelengths and create much higher-quality mammograms.

Figure 2 compares a synchrotron image of a phantom (a physical simulation of tissue) to a traditional X-radiograph. You can clearly see the improved fidelity of this new type of image which offers the radiologist the prospect of earlier detection of malignancy and therefore earlier diag-

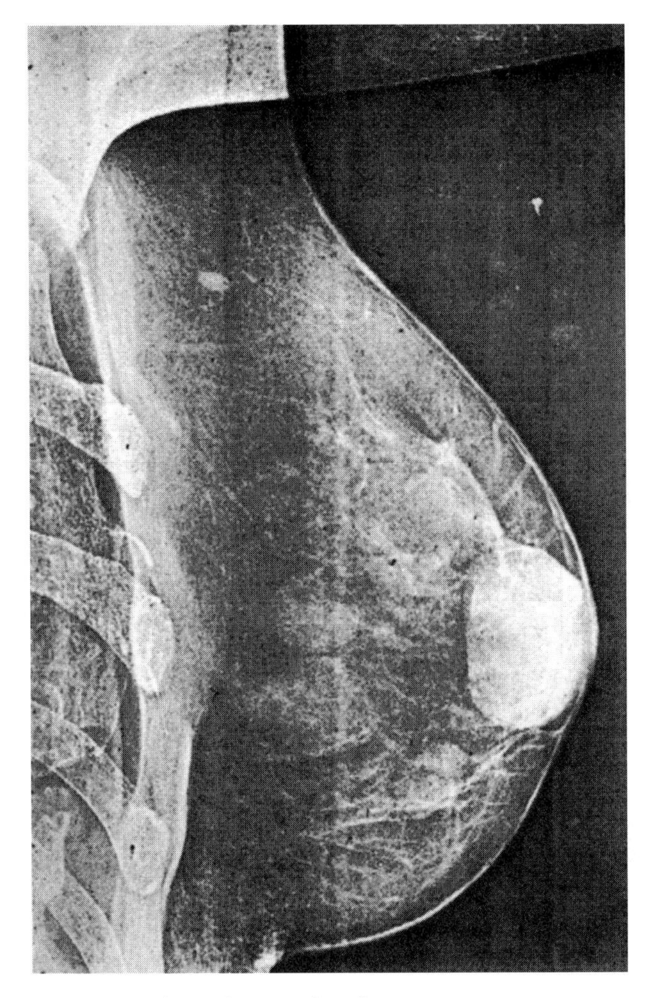

Fig. 1 Conventional mammogram

nosis—the key to improved prognosis. These synchrotron X-rays are currently produced at the UK Research Councils' Central Laboratory at Daresbury where they are traditionally used to probe the atomic structures of matter by physicists and materials scientists. So one can instantly see the 'spin-off' from this 'big' hard materials science into the health-care sector. Currently it is not only the science that is 'big' but also the synchrotron source itself, which is the size of a factory. So there will not be a request for voluntary subscriptions to buy one for your local hospital in the near future. However, a synchrotron is not the only possible source for such X-rays and development work is under way to shrink the size of these sources to make them available for everyday use in larger hospitals.

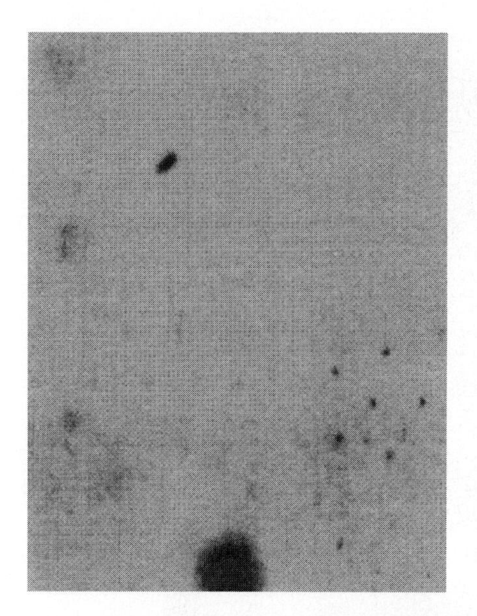

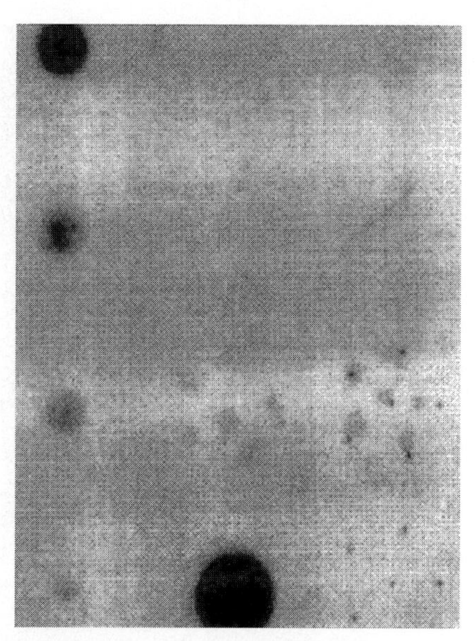

Fig. 2 Normal radiograph (left) and synchrotron radiograph (right)

But this is not the only prospect synchrotron X-rays offer us in the diagnosis of breast cancer. I think it will interest members of the Royal Institution to know that the work of the Braggs, for so long connected with the RI, is also playing a major role in our work on breast cancer diagnosis. Their research on the determination of the atomic structures of solids by X-ray diffraction forms the backbone of analytical techniques in solid-state physics and materials science. This is now being used by our team as a possible new technique for the early *in vivo* diagnosis of breast cancer. X-ray diffraction can characterize the atomic structure of a solid by analysing the X-rays which are characteristically 'scattered' by the regular arrays of atoms within it. We are using synchrotron X-rays to also characterize the structure of tissue within breasts and breast tumours using diffraction effects at low scattering angles. Information of this type is shown in Fig. 3 for collagen in non-malignant and malignant breast tissue.

As this figure shows, there are very large differences between malignant and non-malignant tissues with the 'ring-like' diffraction data, characteristic of normal breast collagen, disappearing when the tissue becomes malignant. In other words we have the prospect of a new and extremely sensitive *in vivo* diagnostic technique based directly on the pioneering work of the Braggs in materials science. This new structural information is available from the X-rays which are normally discarded from conventioinal mammograms and can therefore be collected simul-

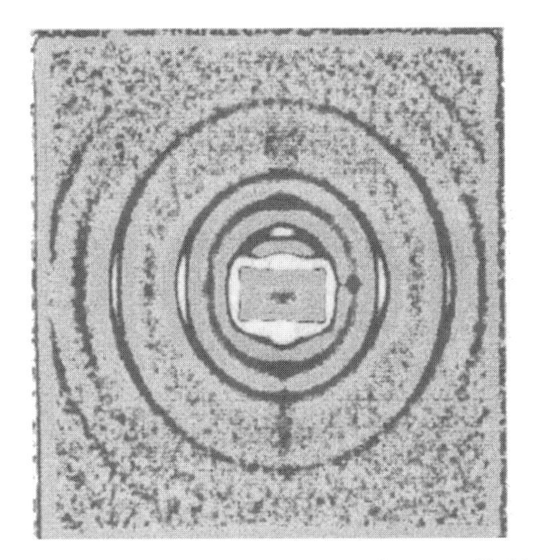

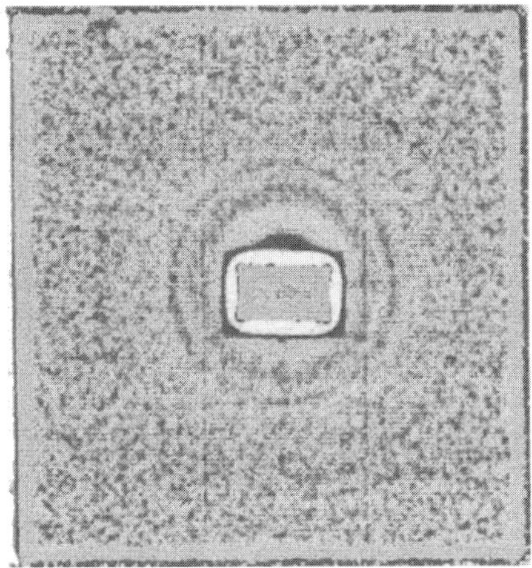

Fig. 3 Normal tissue (left) and cancer tissue (right)

taneously with the new high-fidelity mammograms. This offers the prospect of a dual-edged diagnostic tool: enhanced mammograms with a diffraction-based diagnostic probe—a powerful new weapon in the fight against breast cancer.

I have illustrated an example of the use of materials-science-based analytical techniques to enhance diagnosis in the health-care sector, but the role of materials science need not be limited to diagnosis. 'Life-saving' materials science also stretches to physical materials for health-care. We are working on such novel materials as well. What enables us to do this is our high-intensity cobalt radiation source. What clever science does this allow us to do? Well, it enables us to carry out exciting chemistry involving radiation as well as conventional chemical reactions. Such 'radiation chemistry' can produce materials with interesting functional properties for the health-care field. We can take simple polymers (plastics) from polythene to more specialist materials (such as EVA—ethylene vinyl alcohol—for health-care products) and carry out chemical reactions under the influence of ionizing radiation to produce graft co-polymerized materials (Fig. 4): for example, by adding species derived from polyacrylic acid.

The chemistry is clever, but perhaps even cleverer are the resulting products. For example, these co-polymers enable us to create drug reservoirs for transdermal drug delivery systems such as nicotine patches.

Perhaps of more interest, however, is the wider range of drugs and therapeutic inoculations which can also be delivered from such materials, ranging from familiar therapies such as ibuprofen for pain relief to the nerve-agent pretreatments we have heard so much about in the press related to Gulf War syndrome.

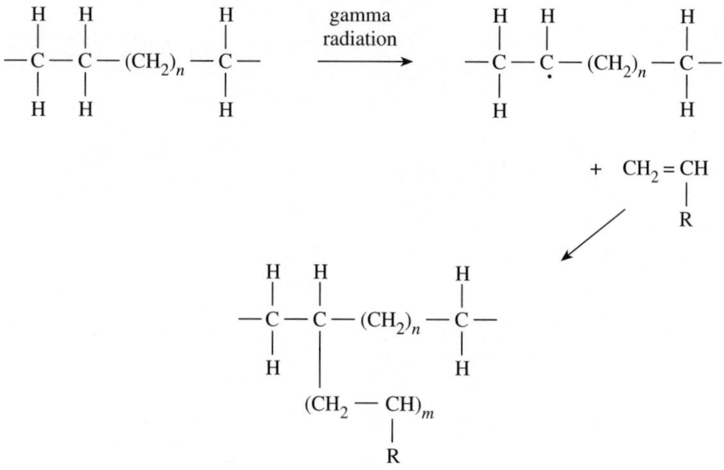

$$-\mathrm{CH_2}-\mathrm{CH_2}-(\mathrm{CH_2})_n-\mathrm{CH_2}- \xrightarrow{\text{gamma radiation}} -\mathrm{CH_2}-\dot{\mathrm{C}}\mathrm{H}-(\mathrm{CH_2})_n-\mathrm{CH_2}-$$

$$+\ \mathrm{CH_2}=\mathrm{CH}(\mathrm{R}) \longrightarrow -\mathrm{CH_2}-\mathrm{CH}\big[(\mathrm{CH_2}-\mathrm{CH(R)})_m\big]-(\mathrm{CH_2})_n-\mathrm{CH_2}-$$

Fig. 4 Radiation graft co-polymerization

The same materials technologies can also be tuned to create families of hydrophilic polymers which strongly absorb water (Fig. 5). These will swell in water-rich environments as well as produce low-friction surfaces. Figure 5 shows an oesophageal stent produced using this technology which can be delivered down the throat when the oesophagus needs to be held open due to disease or other medical interventions. Such a stent can be delivered unswollen and then flushed with warm saline to

Fig. 5 Hydrophilic oesophageal stent

make it expand, providing *in situ* stenting. These effects are all derived from the ability of this technology to control the transport of chemicals such as water through the polymer. This generic technology can therefore be used in stenting ranging from the bile duct to urinary tract, as well as in novel wound dressings, such as those for burns.

I hope I have convinced you in these few examples that materials science is not simply about bridges, aircraft, and other large engineering structures, but also offers us prospects for interesting new products and techniques in the health-care sector. We must now leave 'Life saving' and move on to the role of materials in improving the quality of life and the environment.

Quality of 'life'

If one has a cobalt source and skills in radiation chemistry one can develop products not only for the health-care sector but also to enhance our quality of life. Environmentally friendly energy such as high-powered electrical batteries is such an application. These too depend on the use of interesting graft co-polymers as separators which enable the transport of reactants from one part of the battery to another. However, perhaps more interesting for environmentally friendly power generation is the use of these polymers in fuel cells.

These power sources generate electrical power directly from fuels without burning them. At the core of this technology are solid polymer electrolytes and it will now be no surprise to hear that radiation graft co-polymers offer one route to such solid electrolytes. I have tried to link the health-care applications of graft co-polymers to those in environmentally friendly power generation to show how many of these materials technologies are generic and not sector-specific at all. Different novel materials technologies can therefore impact on our everyday lives across a very wide range of applications.

The impact of materials technologies on environmentally friendly energy need not restrict itself to fuel cells. What about solar energy? How can materials technologies enhance our ability to capture sunlight to generate electrical power? Solar cells have been around for a long time. Based on silicon technology they are still costly and not terribly efficient. What can be done to improve the current situation? The key is to seek new solar cell technologies. Solar cells are essentially semiconductor devices consisting of the p–n junctions familiar from silicon chip technologies.

Figure 6 schematically illustrates current silicon technology which requires relatively large amounts of material compared to the thin-film cells which would be cheap and optimal for fixing to the exteriors of

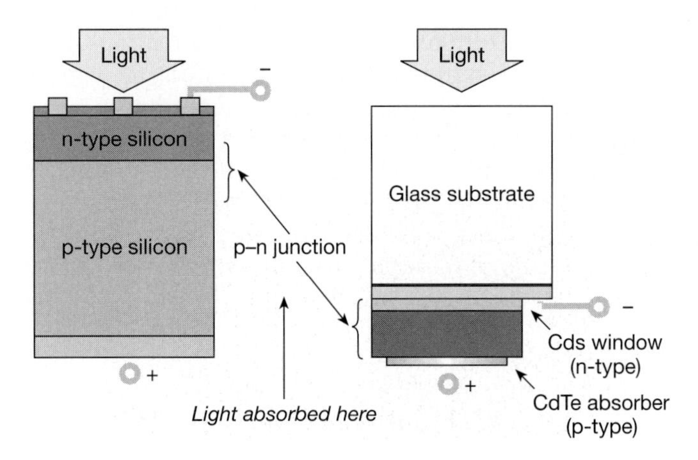

Fig. 6 Schematic diagram of conventional Si and newer thin-film solar cells

buildings. Thin-film solar cells could be coated onto the glass fascias of buildings, generating large amounts of power despite the relatively low intrinsic efficiency of the cells. One of the new technologies is CdTe/CdS thin-film cells (Fig. 6) which are able to be coated onto glass and are cells on which we are working hard to develop efficiencies of around 15% to compete with the best silicon devices. Optimization of these new cell materials requires extensive materials engineering. For example, during processing, CdTe deposits in the incorrect semiconductor form and must be heat-treated to create an active solar cell. However, during such treatments sulphur atoms in the CdS diffuse around the materials destroying the efficiency of the cells—Catch 22! It is surprising that such a phenomenon has not been studied before, however, existing information on the diffusion of sulphur in these metals is extremely limited despite its impact on the future of solar-cell materials. We are studying such diffusion effects from both experimental and modelling viewpoints, learning how CdTe and CdS interdiffuse during heat treatment so that the process can be optimized to produce the appropriate internal structure within the cells and therefore the highest efficiencies.

Figure 7 illustrates the results of computer modelling of these systems, showing the interdiffusion that takes place due to sulphur migration during heat treatment. Modelling is an expanding tool for the materials scientist, enabling us to predict accurately how the materials will perform, allowing us in this case to optimize a system of great significance for the future of environmentally friendly power generation.

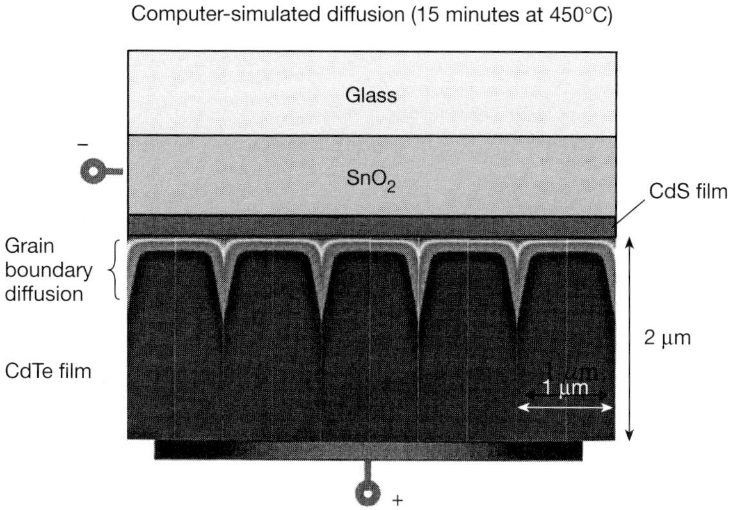

Fig. 7 Computer simulation of the interdiffusion of CdTe and CdS during heat treatment at 450°C

'Life' style

We have seen that many of our new interesting materials are moving away from traditional applications, but you've not seen anything yet! There are plenty of unusual and 'wacky' functional materials we have yet to use in our everyday lives, but they are coming. 'Life style' is becoming increasingly important with designer labels dominating many of our clothes and consumer goods. How can unusual materials contribute to such changes in life style? Let's look first at some unusual materials technologies and then we will see how they will affect our everyday lives.

The first materials I'd like to consider are those which spontaneously change shape on demand. We have seen hydrophilic materials which do this in water but there are other materials technologies such as shape-memory alloys and polymers which undergo similar shape changes when exposed to heat. The shape-memory effect is shown in Fig. 8.

Thermochromic materials are another technology, familiar to us from reactive packaging, beer mats, and fridge shelves. These also react to heat—this time changing appearance. Figure 9 shows an example of such packaging.

Finally, let's consider electrorheological fluids—perhaps a less familiar technology. In the case of these materials, an electric (or, in the related magnetorheological fluids,) magnetic field converts a free-flowing liquid into a soft solid reminiscent of a wine gum in stiffness.

Fig. 8 Shape-memory alloy which will spontaneously move between two memorized 'shapes' on the application of heat

These three materials illustrate a move away from 'structural' engineering materials to alternative functions. But what can one do with such 'reactive' technologies? We are currently receiving significant interest in our work on reactive technologies for domestic products—'humanware', that is, products which interact with their users. Such products are around us all the time but we may not understand how important humanware is and, more importantly, how these reactive materials technologies will help optimize these goods. Let's first take a brief journey into the world of industrial design before we return to 'smart' adaptive products.

It is difficult to believe how we are 'manipulated' psychologically by industrial designers. We purchase items manufactured on a production line in a factory but still consider their 'craftsmanship'. 'What "quality"', we say, and desire to collect more of their type, even falling in love with our favourite products—think only of the love affairs many of us have with products such as cars. Industrial designers deliberately create this desire for products, often without our even knowing. Currently they do

Fig. 9 Thermochromic heat-reactive packaging

this mainly using visual form; however, think of the possibilities of products reacting to us in real time using the reactive technologies we have already discussed. Think about the possibilities for generating 'pet appeal'—the design feature which makes us 'fall in love' with products. The possibilities are endless. But the use of such materials technologies in product design need not end simply in creating new product aesthetics. They will also play a major role in improving the function of future humanware.

How many of us fail to use products correctly, from displays and buttons on consumer electronics to car controls and door furniture. We often think this is due to our own inadequacies, however, in practice it is more often due to poor 'interface' design, the product failing to communicate its function to the user. This is a failure in the product's 'affordances', the visual form which subconsciously communicates the function of the product and how to interact with it. Correct affordances can be created by conventional visual design, but think of the potential of 'dynamic affordances', products which react in real time to their function by shape changes, colour changes, or changes in surface texture or compliance. The prospects for such a new generation of improved products are enormous. So what are we doing to create new dynamic affordances in humanware? Much of our product work is commercially confidential since it is so close to market. However, we can illustrate this design concept with some widely publicized work we have carried out on thermochromic reactive technologies in domestic appliances such as kettles; these communicate functions such as heating fill level through a dynamic visual aesthetic rather than through the electronic components and displays we currently use. Simple physical mock-ups of such products exist but perhaps more interesting is our computer modelling in this area, some of which is shown in Fig. 10.

Modelling allows the full effects of these new design paradigms to be explored during concept development and it can also be used to assess consumer preference using virtual reality.

In conclusion

We have reached the end of our journey for now and I hope I have convinced you of the importance of materials technologies and the energy and enthusiasm of those exploring the 'cutting edge' of materials research. Our science is relatively new and many of the applications in

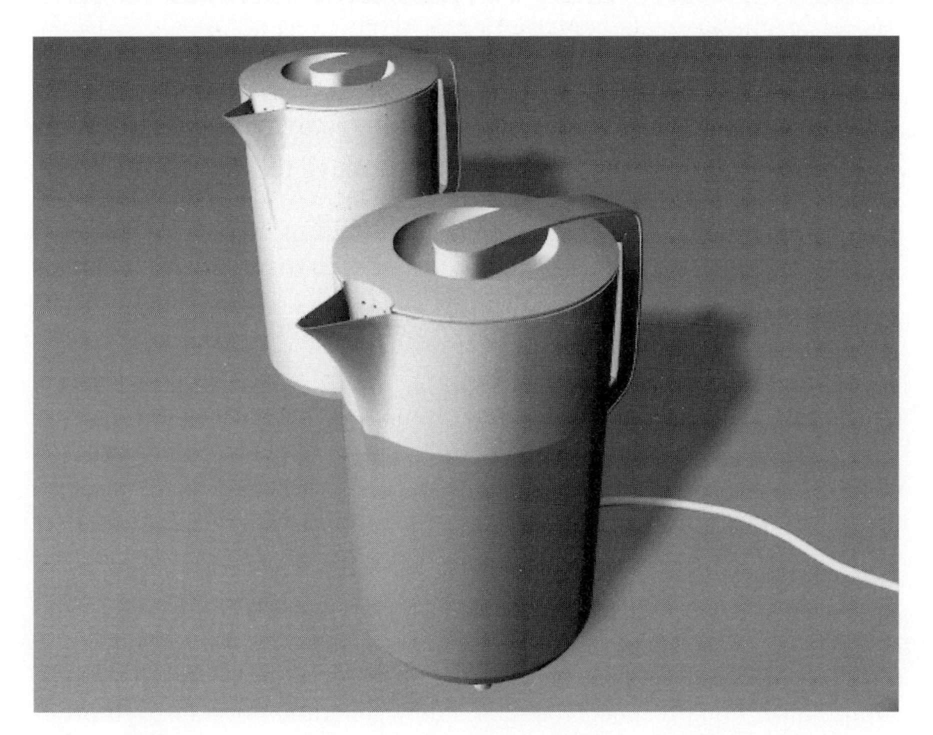

Fig. 10 Dynamic affordances available from a 'colour-changing' kettle

'less familiar markets' are even newer. I hope I have also convinced you that materials have impacted on our everyday lives for thousands of years and continue to do so more and more as our lives become more technology-intensive. Many of these materials technologies are and will become increasingly familiar in our everyday lives in sectors ranging from health-care to 'green' power and humanware products.

It only remains to acknowledge my close working colleagues without whom all this work would not have been possible, including Dr Keith Rogers, Keith Lovell, Dr David Lane, and Christopher Thorpe. To these gentlemen I extend my warmest thanks for their enthusiasm, technical insight, friendship, and continuing commitment to research on materials for the new millennium.

CLIFFORD. FRIEND

Born in 1959, he was educated at the Forest School, Winnersh, Berkshire and the University of Surrey. He obtained a first-class honours degree in metallurgy in 1981, followed by a PhD in 1985 for pioneering work on shape-memory alloys. He has held academic posts at Cranfield University since 1985, latterly as Head of Department and Professor of Materials and Medical Sciences. He has active research interests in composite materials, shape-memory alloys, biomedical materials and devices and smart materials and structures, and has held visiting posts in Göteborg, Sweden and Barcelona. He is a member of many technical committees and editorial boards and was the recipient of the Institute of Materials' HollidayPrize in 1994. His expertise is frequently called upon to assist commercial clients, professional institutions and government agencies ranging from EPSRC to the DTI. He has played an active role in public understanding of science events, contributing to SET Week programmes, the RI/SmithKline Beecham Technology Masterclasses and projects at the National Museum for Science and Industry. He has also acted as an expert for radio, television and many broadsheet newspapers and magazines, including *New Scientist* and *The Financial Times*. He is a Director of Cranfield Polymers and Cranfield Holdings (Wessex) and is retained as a consultant by Ove Arup consulting engineers.

Every drop to drink

KENNETH IVES

Introduction

Two hundred years ago Samuel Taylor Coleridge compiled his lengthy poem 'The Rime of the Ancient Mariner', which described the distress of a sailing ship and its mariners, marooned in the doldrums of the ocean, with the well-known quotation:

Water, water, every where,
Nor any drop to drink.

What the Ancient Mariner needed was a reverse-osmosis desalination plant, but that would not emerge until about 150 years later!

Today we are going to consider our present circumstances in Britain, where water is delivered to our homes with every drop to drink.

It is desirable to link a lecture at the Royal Institution with Michael Faraday. Such a link, albeit tenuous, is possible with the *Punch* cartoon of 1855, depicting Michael Faraday offering his visiting card to Father Thames (Fig. 1). Alas, there is no evidence that he ever engaged in studies on water quality, neither is there any record of consultation between the Victorian Thames Water Conservators and Faraday. There is in the archives a letter to Faraday from one of the water companies of London, asking for his help to eliminate a nuisance organism called 'Conferva' that was growing in the water. It is surmised that this was the filamentous alga *Cladophora*, known colloquially as 'blanket weed'. This is still a nuisance in waterworks filter beds, ponds, and aquaria, although it has no health significance. No reply to that letter was recorded.

A more recent cartoon by Dunn, published in the *New Yorker* in 1965 (Fig. 2) shows a harassed chemistry teacher defining drinking water to his class. In spite of the exaggerations, the message is clear—there were many things in the drinking water at that time which we would rather be without. But that was a third of a century ago.

Fig. 1 Michael Faraday presenting his visiting card to Father Thames.

Quality standards

In the UK and the European Union, every drop must be fit to drink at the household tap (not just at the point of leaving the waterworks). It is remarkable that we only drink about two per cent, directly as tea, coffee, etc.; the rest is used for washing ourselves, the laundry, dishwashing, and most significantly, flushing the toilet. In these latter uses, the quality requirements would not be necessary to be so stringent, except that the regulations require that every drop entering the house must be fit to drink, whatever its subsequent household use. More than sixty Quality Standards are defined in the regulations: some are physical, some chemical, and some microbiological.

An extremely exacting standard is for individual pesticides, which is set to not exceed 0.1 μg per litre, which is equal to one part in ten thousand million (1 in 10 000 000 000 or, in scientific terms, 10^{-10}). On the other hand the standard for sulphate, which can have a laxative effect, is

"Now, when we take three hundred millilitres of a compound containing hydrogen and oxygen in a ratio of two to one and add three millilitres of an eight-hundredths-per-cent chlorine solution, one millilitre of a three-ten-thousandths-per-cent stannous-fluoride solution, and fifty millilitres of treated industrial wastes and solids, we get drinking water." *Drawing by Alan Dunn: © 1965 The New Yorker Magazine, Inc.*

Fig. 2 A harassed chemistry teacher defining drinking water to his class.

250 parts per million (1 in 4000 or 2.5×10^{-4}). These standards are advised by the World Health Organization, based on health risks over a lifetime's exposure. The standards adopted as regulations in EU and UK law require that drinking water must conform. If any of the regulations are breached, the water is not fit to drink, and it would be a statutory offence to supply it, which would be a criminal act. In England and Wales such statutory control is in the hands of the Governmental Drinking Water Inspectorate which regularly monitors the quality of water supply and can prosecute water companies that do not comply.

We, as consumers of water, are unaware of the presence of most of the substances which are limited by the regulations, and sophisticated analyses are necessary to detect their presence and concentrations.

Sensory perception

Exceptions to the need for analytical equipment to determine the presence of offensive substances in water, are those detected by our sensory perceptions. These are:

(1) Turbidity (cloudiness)

Turbidity is measured in Formazin units (FTU), which are chemically reproducible haze standards that water chemists have adopted from the brewing industry. A value of 1 FTU would be just about detectable in a glass of water by the human eye, particularly if the glass were held up to the light. The River Thames at its worst during flood conditions has a turbidity of about 500 FTU. Drinking water requirements are below 1 FTU, and require instruments to detect and measure the turbidity. Most instruments depend on the phenomenon of Tyndall light scattering, which is called 'nephelometry'. We are most familiar with this phenomenon when it is evinced by a bright beam of light (sunlight or a projector beam) shining through air containing fine dust particles. We see the particles by looking at right angles to the beam, because they scatter light out of the beam. Without the beam of light, the dust particles are not visible in normal ambient illumination. In water, nephelometry is very sensitive, detecting concentrations of turbid particles down to one-hundredth of the human eye limit.

(2) Colour

Water may be free of turbidity, but still not clear due to the presence of colour. Colour is primarily due to dissolved, or very finely divided, organic matter in the water, usually providing a light brown tinge. It is characteristic of water from upland regions due to the presence of peaty material, described generally as 'humic acids'. Colour is not detrimental to health, but when combined with chlorine can form substances called trihalomethanes (THMs), which can be carcinogenic, and of which chloroform is the best-known example. Present standards limit this to 100 μg per litre, being one part in ten million (1 in 10 000 000 or 10^{-7}). Colour itself is limited to 20 mg per litre (20 parts per million) measured on an artificial scale prepared from a cobalt-platinic acid solution, often quoted as degrees Hazen (°H), named after the first water engineer to devise a scale of measurement for colour about 100 years ago. Colours below 5 mg per litre are scarcely detected by the human eye.

(3) Taste

The presence of dissolved material in water may be offensive to our palates and we would not drink the water. Human responses to tastes vary with the sensitivity of the consumer, and often intensity and description can be extremely confusing. In a water pollution incident on the River Severn in 1994, water consumers in the city of Worcester complained variously about tastes and odours which were described as 'sweet', 'sewage', 'thinners', 'paint', 'bad eggs', and 'manure'. Yet these were all due to the same chemical which had been released upstream from an industrial site. The chemical was found to be 2-ethyl-5,5-dimethyl-1,3-dioxane (abbreviated to 2EDD), an organic chemical previously unknown in the UK. Because of this variability of perception and the vast range of potential causes there are no direct standards. But a measure of intensity is called the 'dilution number', which is the number of times that the water being tested has to be diluted with taste-free water (usually distilled), for the taste to be undetectable. Standards typically require the dilution number to be less than three.

The perception of taste can be easily tested among a small group of people by asking them individually to identify unlabelled samples of water, for example: tapwater, distilled, bottled (not gassy), and tapwater with a drop of the mouthrinse/disinfectant TCP. It is rare for anyone to identify all four correctly, although most people readily detect the TCP sample.

(4) Odour

This is usually related to the presence of tastes, as we connect taste and odour together. The remarks above about taste apply also to odour. Many odours in water come from algae, particularly when the algae are concentrated by treatment processes in the waterworks. Most of these odours are characteristic of particular algal species. There exist tabulated identifications of odours in water, with their associated species of algae. One such, for example, which has been experienced in waterworks in London is the algal diatom *Asterionella* causing an odour of geraniums. This odour may be pleasant in a garden, but undesirable in drinking water.

These properties of water detected by our senses are called 'organoleptic'.

Hardness

Another property of water, which we can detect, but not measure easily, is its hardness. Hardness is due to the dissolved salts of calcium and magnesium, normally derived from chalky or limestone areas, affecting well and borehole waters as well as springs and streams feeding rivers and reservoirs. The dissolved calcium and magnesium react with soap in the water to form precipitates which make the water cloudy, and reduce the amount of soapy bubbles. By shaking the water with a solution of soap and noting the persistence (or lack) of bubbles, the presence of hardness can be demonstrated.

Hardness does not affect the potability of water, although there are claims that hardness (or lack of it) can be tasted. There are statistical links that show that living in a hard water area lowers the probability of heart disease and circulatory problems. But no causal link has been established between the protection from heart disease, and calcium and magnesium in the drinking water.

Illness from water

If illness is attributable to water, it is usually as intestinal gastric upsets, and these are most likely caused by pathogenic micro-organisms from water contaminated by humans or animals.

Historically, this was established by Dr John Snow (Fig. 3) in 1854 by his investigation of the Broad Street pump in Soho in London. At this

Fig. 3 Dr John Snow, who discovered the link between water and cholera: the first water epidemiologist.

time cholera raged in London, and Soho was particularly menaced by this deadly disease. Medical opinion at the time took in the view that these (and other) epidemics are spread by vapours: the miasmatic theory of disease. Snow plotted on a map of Soho the locations of deaths from cholera, and noted that they clustered around the pump in Broad Street (Fig. 4). Furthermore, workers in a nearby brewery in Golden Square did not die of cholera, because their habit was to drink beer, not water. The association of cholera and water was confirmed in Snow's mind when he learned of the death from cholera of a lady in Hampstead, which was then a village outside London. She regularly sent her servant to collect water from the Broad Street pump, because she liked its taste. No other cholera deaths occurred in Hampstead at that time. This was the dawn of water epidemiology, although the germ theory of disease and the identity of the pathogenic organism *Vibrio cholerae* were not established until some decades later. John Snow was already a notable gynaecologist, and administered anaesthesia to Queen Victoria in childbirth. Today he is honoured both as an anaesthetist, and an epidemiologist, through the John Snow Society which holds regular meetings to commemorate his activities. A further commemoration exists as the John Snow public house (Fig. 5) in Broadwick Street, Soho, reputedly on the site of the original pump. Also, a mock Victorian pump has been set up

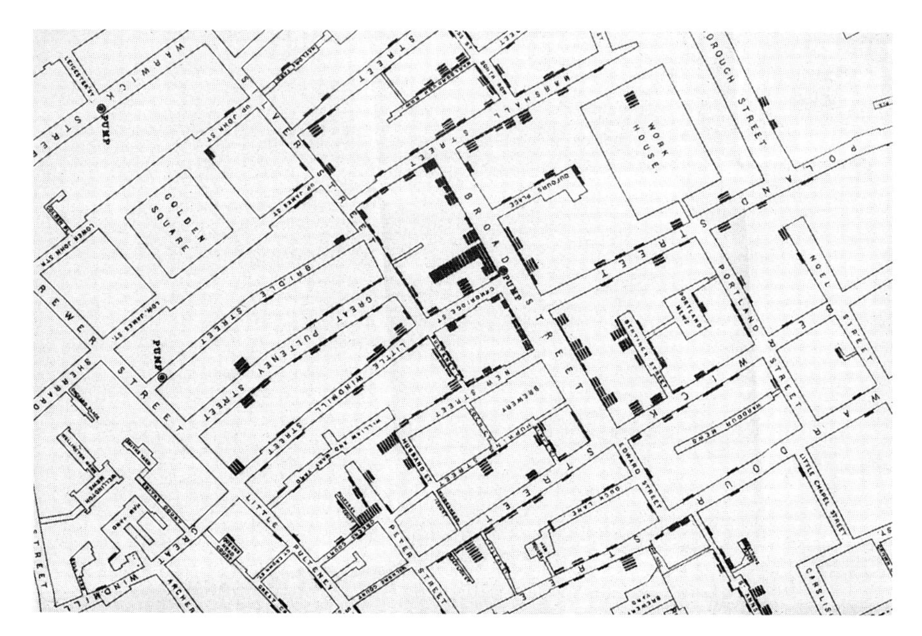

Fig. 4 Snow's map of Soho, marking the houses with cholera deaths, which cluster round the Broad Street pump.

Fig. 5 The John Snow Public House on the site of the Broad Street pump.

a few metres away, to remind passers-by that John Snow's advice for dispelling the threat of cholera was to 'take the handle off the Broad Street pump'.

Now, pathogenic micro-organisms in water, including bacteria, viruses, and some protozoa, are killed by adding chlorine, and occasionally ozone. Chlorine is highly effective, cheap, and easily administered, but its use has diminished a little, particularly in some continental European countries due to potential formation of carcinogenic by-products. (See previous remarks about THMs.)

Cryptosporidium

This dissatisfaction with chlorine as a water disinfectant has been reinforced by the detection of the pathogenic protozoan *Cryptosporidium parvum* (Fig. 6) in water supplies, and some consequent epidemics of cryptosporidiosis, which is an acute and distressing disease causing diarrhoea and vomiting. It is not fatal in healthy adults, but can be life-threatening in immuno-compromised patients. The infective form, called an 'oocyst' (Fig. 6) is widespread in the environment and may be derived from farm animals and wild animals as well as from cryptosporidiosis sufferers. It has been detected in the water in several countries, and has caused serious outbreaks in the USA, the UK, and Japan.

Oocysts of *Cryptosporidium* are resistant to chlorine, and have variable resistance to ozone, and so chemical disinfectant of water is being replaced in several locations by improved methods of physically removing them. They are microscopically small, about 4 μm (four-thousandths of a millimetre) in diameter, so enhanced filtration through sand, or the use of fine-pored membrane, is gaining approval to meet this new menace in the water.

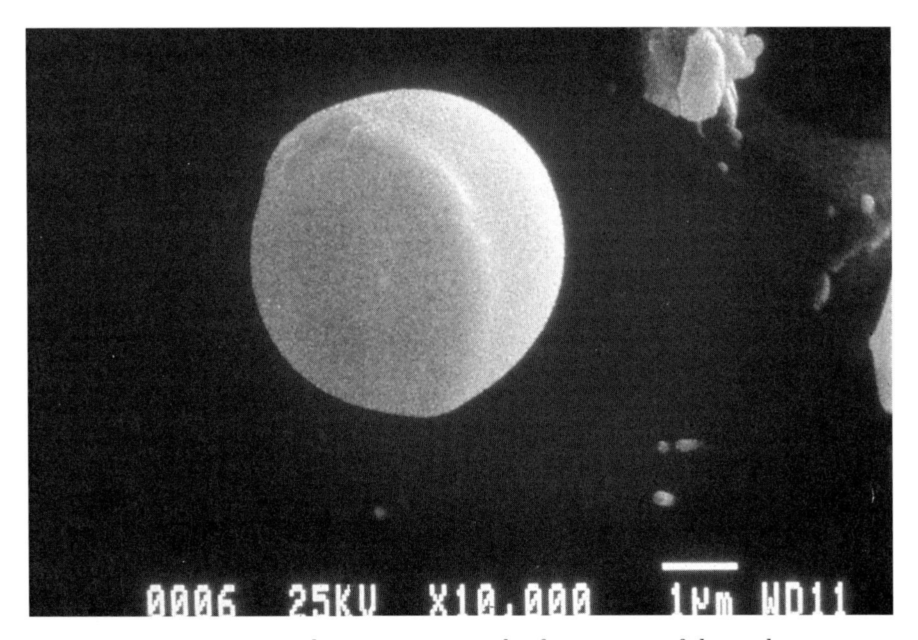

Fig. 6 Scanning electron micrograph of an oocyst of the pathogenic protozoan *Cryptosporidium parvum*. It is about 4 micrometres in diameter.

Slow sand filtration

The Fall 1997 issue of *Life* magazine's special double issue was devoted to '100 events that changed the world' in the last millennium. Event No. 46 was headed '1829 Water Purification', and contained the following: 'But before 1829, when the Chelsea Water Works of London installed its landmark slow sand filter on the Thames River, no one had effectively cleaned it.' This was a tribute to the pioneering concept of filtration of river water through sand for public supply, by the water engineer James Simpson. His reasoning was that water that had percolated underground became clear and sparkling when taken from wells, or emerging from springs.

(Incidentally none of Michael Faraday's inventions or developments appeared in *Life*'s 100 events.)

Slow sand filters are still used in London, but the originals disappeared when the Chelsea Water Works site was built over by Victoria Railway Station. They are to be found at water treatment works in the Thames Valley, and most recently at Coppermills Water Treatment Works in Walthamstow, all feeding the ring main deep under London. Coppermills Works, shown in Fig. 7 comprises 32 slow sand filters, each about the size of a football pitch, and produces about 600 megalitres of

Fig. 7 Aerial view of Coppermills Water Treatment Works (Thames Water Company) at Walthamstow in London. The slow sand filters can be clearly seen.

drinking water (600 000 tonnes) per day. This is sufficient to supply about one hundred thousand households per day.

This illustrates the vast scale of engineering the supply of drinking water to the public, contrasting with the minute scale of impurities that have to be removed—see the earlier comments on quality standards. It is a challenge that has to be met every day, for 24 hours per day for 365 days per year, at a suitable pressure for every home.

Slow sand filters have advantages which were unforeseen by James Simpson: his objectives were aesthetic, to provide attractive drinking water to the customers of the Chelsea Water Works Company. His filters led to an improvement in the health of the consumers of water, and the 'English filters' began to be adopted in Europe, the USA, and the British Empire. The reason was that slow sand filters remove more than 99 per cent of the intestinal bacteria from the water. But the link between water and disease was not made until two and a half decades later by Dr John Snow in Soho, and it was later still in the Victorian era that disease bacteria were identified by microscopic and culturing techniques. An outstanding recognition of the value of slow sand filtration occurred in Germany in the twin towns of Hamburg and Altona, both drawing water from the River Elbe. Altona, which had installed the 'English filters', suffered a death rate from cholera of only 2.13 deaths per 1000, whereas Hamburg without the filters had a cholera death rate of 13.39 per 1000, over six times as many.

But slow sand filters are not without their drawbacks. Because most of them in the UK are open to the sky they can develop algal growths of the nuisance organism 'blanket weed' (*Cladophora*) which makes cleaning of the filters very difficult (Figs 8 and 9). It is a paradox that the action of microbiological growths on the sand, which are essential to their proper functioning, can lead to massive overgrowths if the water contains dissolved minerals and is exposed to daylight.

Coagulation–flocculation

An alternative to the biological filtration action of the slow sand filters is a process which aggregates fine particles in water, such as those of turbidity, micro-organisms to a limited extent, and even colour, is called 'coagulation–flocculation'. Because particles in water carry a small electrical charge, they mutually repel one another and so cannot be aggregated without the intervention of certain chemicals called 'coagulants'. These chemicals, of which alum (aluminium sulphate) is typical, reduce the electrical repulsion so that if the particles approach one another in the water they can touch. This process is called 'destabilization', as it

Fig. 8 A close-up of blanket weed (*Cladophora*) on the sand surface of a slow sand filter.

Fig. 9 The piling up of blanket weed before the dirty sand surface can be scraped for cleaning.

removes their continued stability as individual particles. Once they are destabilized, molecular forces causing random movements, known as 'Brownian motion', and hydrodynamic forces can push the particles together so that they stick, and aggregate to sizes that can be removed by non-biological sand filtration, or by settling, or by flotation with air bubbles. This process of aggregation is flocculation and is assisted by chemicals such as alum which form precipitates to enhance the rate of particle collisions. The addition of alum, or similar chemicals, assisted by a slow stirring action, leads to the formation of visible aggregates, called 'flocs', which can then be removed by the physical processes of filtration, settling or flotation. The coagulation–flocculation operation is readily demonstrated in a series of glass jars, each of one litre capacity, fitted with stirrers, as shown on Fig. 10. This is known as a 'jar test' and is used by water chemists to determine the correct chemical conditions for the formation of flocs. But this is only a laboratory mimic of the engineering scale which is necessary in a water treatment works. There the stirrers can be up to three metres in diameter and several metres long, contained in tanks with capacities of several hundred cubic metres of water. (Note that one cubic metre of water weighs one tonne.) They create visible flocs, a few millimetres in size, which contain the impurities which were in the water. These flocs are separated from the water by settling, flotation, and filtration.

Fig. 10 A laboratory jar test.

Separation processes

Settling

The processes of settling rely on gravity to remove flocs from the water as a sludge. This is a slow process, because the flocs are only slightly more dense than water. Consequently the settling tanks have to be very large to give the flocs enough time for their settlement while the water is in the tank. Such a tank is shown in Fig. 11; the long channels are decanting the clarified water, to pass it to filtration for final cleaning.

Fig. 11 A horizontal flow settling tank. Flow is away from the camera. A cloudy suspension of flocs can be seen in the foreground. The channels which collect the clarified water are in the background.

Flotation

A more recent process, developed over the last two to three decades, relies on the release of fine bubbles from water under pressure; these collect the flocs and float them to the surface, where they are skimmed off as a froth. Before entering the tank where flotation takes place, part of the water is mixed with air under high pressure so that the air dissolves. As this pressurized stream of water joins the main flow into the flotation tank, the pressure is released and the air appears as a cloud of fine bubbles. It is similar to releasing the cap on a bottle of gassy mineral water and seeing gas bubbles appearing in the bottle. However, in that case the gas is compressed carbon dioxide, which would be much too expensive compared with air for engineering-scale water purification.

As the flocs stick to the bubbles they rise with them at a faster rate compared with gravity settling, and the flotation tanks are consequently smaller. An example of this dissolved-air flotation process is shown in Fig. 12.

Filtration

About 60 years after James Simpson's development of the slow sand filter (see the previous commentary in this paper), a new form of sand

Fig. 12 Dissolved air flotation tank with floc and bubble froth on the water surface.

filter emerged. This uses coarser sand, and the water flows through it 50 times faster than in the Simpson design. This has led to the terms 'slow' sand filter and 'rapid' filter. Whereas the slow sand filter is essentially biological in action, the rapid filter does not retain the water for long enough for the appropriate microbiological growths to occur. Its action is physical and chemical, relying on microscopic physical and hydrodynamic forces to remove flocs from the water, and with them their associated impurities. Because the flow through these filters is so rapid, they occupy a much smaller area than the slow filters. But the rapid rate of flow causes these filters to clog more rapidly with flocs and impurities, and they have to be cleaned every 24 hours, compared with cleaning of slow filters every two or three months.

Figure 13 shows a rapid filter, which is contained in a concrete box-like tank, very much smaller than the football pitch areas of the slow sand filter in Fig. 7. Rapid filters are cleaned by blowing air upwards through the clogged sand, and then flushing the dirt away by an upflow stream of water. This procedure, called 'backwashing', lasts only a few minutes, before the filter is put back into service again. It can be seen in Fig. 14. Recent research has shown, both theoretically and in practice, that the cleaning of the sand grains is at an optimum when air and water are applied simultaneously in a certain proportion, which produces a cleansing action known as 'collapse-pulsing'. Theoretical development of this

Fig. 13 A rapid sand filter, drained down ready for backwashing.

Fig. 14 Rapid sand filter backwashing. Note the dirty wash-water which is leaving the sand underneath.

concept from a basis of soil mechanics and fluid dynamics was made at Georgia Institute of Technology in the USA, and was confirmed experimentally at University College London using endoscopy to view the inside of a filter during its backwashing. The resulting video-recording of collapse-pulsing has been shown at international meetings of water specialists, and the technique is being adopted in engineering designs.

Carbon adsorption

The presence of organic materials in solution in water has led to the strong development of activated carbon adsorption. The organic materials include taste and odour compounds, colour, and microcontaminants such as pesticides and herbicides, and the industrial pollutant 2EDD, which has been mentioned previously. The extremely low limits of organic substances in regulatory standards for drinking water has led to very careful assessment of their adsorption on to activated carbon grains. The process of granular activated carbon (GAC) filtration is now widely adopted in water treatment, sometimes with the addition of ozone which breaks up the organic molecules, rendering them more readily removable by GAC. Sometimes, in emergency water treatment, for example after an accidental spill of organic substances into water supply, powdered activated carbon (PAC) is added to the water. But when the PAC has adsorbed the undesirable chemicals, the powder has to be filtered out, usually in rapid filters. The use of GAC does not require such secondary filtration, but when its capacity to adsorb organic chemicals is exhausted, the granular carbon has to be regenerated with high-temperature steam treatment to restore its highly porous activated state.

Home filters

There is a growing trend in the use of home units to filter tapwater before use. These are called 'point-of-use devices' and typically contain a removable cartridge containing fine granular beads. In such cartridges there is usually granular activated carbon, which can remove tastes, including those due to chlorine, which may be undesirable to the water consumer. In addition, in hard water areas, the cartridges may contain ion-exchange beads which remove calcium from the water, replacing it with sodium. This makes the water softer, thus reducing soap scums, scale in kettles, and calcium scums which may form on tea and coffee.

Such point-of-use devices do not make the water safer, or more healthy to drink. Their value lies in their ability to satisfy the aesthetic response, which varies from consumer to consumer, with respect to appearance and taste. In most designs the cartridge is thrown away when its capacity is exhausted, and a new one must be purchased, which is an extra cost to the consumer. An exhausted cartridge, or nearly exhausted one, may be in fact a detriment to the quality of the water, as carbon grains will collect (adsorb) organic molecules. The accumulation of organics provides a good supply of food on which bacteria may grow and increase, adversely affecting the very qualities which the customer desires.

In hard water areas some householders install permanent softening systems using ion-exchange resin beads, to protect their plumbing from scale formation, as well as supplying soft water for bathing and washing. These permanent installations do not use granular activated carbon, because it is not needed for most household purposes, and it would be difficult to regenerate. Ion-exchange beads for softening, however, can be regenerated in place with common salt solution. Miniature versions of this softening procedure, with salt regeneration, are to be found in domestic dishwashers, to avoid cloudy scale formation on crockery and the inside of the dishwashers.

Bottled water

Connected with the increased use of home filters is the rise in consumption of bottled water, both still and gassy. This owes much to the skills of marketing by bottled water suppliers. But there exists a public perception that bottled waters are healthier than tapwater, and that it is fashionable to drink them. The latter is probably true, although it is a self-fulfilling concept, particularly as restaurants are happy to provide such an overpriced commodity.

As for home filters, the justification for drinking bottled water is primarily aesthetic, people preferring, or becoming accustomed to, its taste. The issue of hardness does not arise, as usually bottled waters contain more dissolved minerals than hard tapwater. Although their contents are now controlled under UK law, the analyses given on the bottle labels rarely mention bacteriological testing. The real issue is the price: tapwater arrives under pressure at about 70 pence per tonne; bottled water has to be fetched, at a price up to £1000 per tonne. Some UK water companies are aware of this market, and sell bottled water in competition with their normal drinking water supplied at the tap.

Don't waste water

Whatever the source, the current message is 'don't waste water'. In times of drought, restrictions are applied to the use of water for various purposes, notably garden watering. This message is being reinforced by the introduction of water meters. In houses of only one or two occupants, the price usually represents significant saving over the previous rateable value costs. If an overall reduction in the use of water results from universal metering, the present metered price cannot be retained: the overall revenue to the water company must be maintained, or even increased, and so the price of water will go up per tonne delivered.

A variation on the standard price per tonne (cubic metre) irrespective of the quantity used, is to charge more per tonne for higher use. On a per capita basis a quantity is set (for example per month) which meets the basic health and hygiene requirements: possibly about 150 litres per day per person. Demand beyond that limitation would be deemed luxury use, and attract a higher and higher cost per tonne. Such a system is in place in South Australia, for example, which is a water-short state where water saving is encouraged.

This apparently modern concern for the careful use of water is not new. Just over 100 years ago, a cartoon in *Punch* in 1896 showed an east London turncock addressing a group of women with jugs standing around a newly installed standpipe tap, with the admonition: 'Now, look 'ere, don't you go a-wasting all this 'ere valuable water in washin' and

Fig. 15 East London turncock admonishing women not to waste water.

waterin' your gardens, or any nonsense of that sort, or you'll get yourselves into trouble' (Fig. 15)

Back to the Ancient Mariner

When the calm broke, and the ship reached land and fresh water

Gramercy! they for joy did grin,
And all at once their breath drew in,
As they were drinking all.

A happy ending, which we all wish ourselves.

KENNETH IVES

After graduating as a civil engineer, Ken Ives was employed for seven years with the London Metropolitan Water Board (now Thames Water), working principally on water treatment. Research on water purification led to a PhD at University College London (UCL), followed by an academic post there. A period of postdoctoral study at Harvard established a special interest in water filtration. This continued at UCL, and a subsequent US National Science Foundation Senior Fellowship at the University of North Carolina consolidated further research interests in flocculation. He was a Visiting Professor at Delft Technical University, and spent 25 years as an Adviser on Environmental Health to the World Health Organisation. Recently he served on the UK Government's Panel of Experts on Cryptosporidium and Water Supplies (Badenoch Committee), and led the Inquiry into the water supply incident in Worcester, and the diesel spillage problem in Edinburgh's drinking water.

Mirrors in mind—revisited[1]

RICHARD GREGORY

Reflections have intrigued the curious throughout recorded history, and no doubt from long before the Egyptian metal mirrors of perhaps 5000 years ago. The first personal mirrors were water carried in bowls, but these were inconvenient for showing more than the down-turned face. Water was sometimes turned on its side, by wetting slate hung on a wall. Then there were mirrors of burnished copper, gold, and, most precious and best for the face, silver. Egyptian mirrors were considered to be windows into another world, and were buried with the dead.

For the Greeks, reflections deep in lakes were seen as too dangerously fascinating, perhaps because one's double, drowned yet almost alive, is the *Doppelgänger* of horror, haunting by mocking our passing time from its shadow-world of intangible gods and ghosts. Mirrors are frightening, magical, and puzzling.

Puzzling reflections

From physics to physiognomy—with some psychology thrown in—reflections in a looking-glass have confused even philosophers and scientists for millennia. Some mirror puzzles are simply answered with optical ray diagrams. Why must a mirror be at least half one's height and breadth to show one completely from any distance? Why is the image as far behind as the reflected object is in front? Why does a tracing of one's head on the glass continue to be filled with your head when viewed from however far away? Although perhaps surprising, these are answered by simple optics with ray diagrams. But why does a looking-glass reverse horizontally, but not also vertically? This may look like the same sort of question as the others; but perhaps surprisingly it cannot be answered with a ray diagram &. What kind of answer, then, would be appropriate? This is an ancient controversy going back to Plato. Before looking at this we will, however, glance at reflection itself.

How do mirrors reflect?

How mirrors reflect light is not obvious, and has become more surprising as science has advanced. Why light reflects the way it does was discussed by the Greeks, who studied mirrors as a branch of science. They knew that the *angle of reflection equals the angle of incidence.* They compared this with how a ball bounces off a surface; but it became more mysterious when this 'obvious' notion had to be abandoned. For, in the nineteenth century, it was found that the material of the mirror matters. So light must 'see' into the material of the mirror, rather than simply bouncing off its surface. What showed this, was the observation that Brewster's angle of maximum polarization depends on the refractive index of the reflecting material.[2] It turns out that the wider generalization of the law of reflection, is that light takes the *shortest time-path.* For mirrors, this is the same as the shortest distance, as the light is travelling at constant speed; however, for refraction the shortest distance is not the shortest time-path.[3]

It turns out that as light 'sees' into the mirror, it activates a dance of electrons in its material. Choreographed by principles of quantum electro-dynamics (QED), it is re-emitted as fresh photons. But why should it re-emit at the same angle that the light entered? How can it select the place on the mirror for the shortest time-path? We can see where this place is, by simple geometry (Fig. 1).

Explaining this, needs an imaginative world far from familiar experience or intuitions. The QED answer starts by denying that it is only this one region of the mirror that matters. Rather, all parts of the mirror contribute. It is the strange nature of light itself that makes the central region most important. Ray diagrams cannot show why light takes the shortest path, and this is only statistically true; for photons may take *any allowable path,* though some are more probable than others. Here I shall draw on Richard Feynman's fascinating *QED: The Strange Theory of Light and Matter* (1985). My account here will be inadequate, but although this is not our main theme it seems right to consider the current explanation, though only briefly. Here we dip below familiar observations to a hypothetical world—to emerge to familiar and yet puzzling everyday reversals of mirrors.

QED

When light is very dim, individual photons can be detected. However feeble the light, each detected event has full strength, as for bright light, though the events are rare. So light is seen to be particles—*photons.* At higher intensities light may be described as waves; but even at low intensities, when there are but few particles, each photon has wave proper-

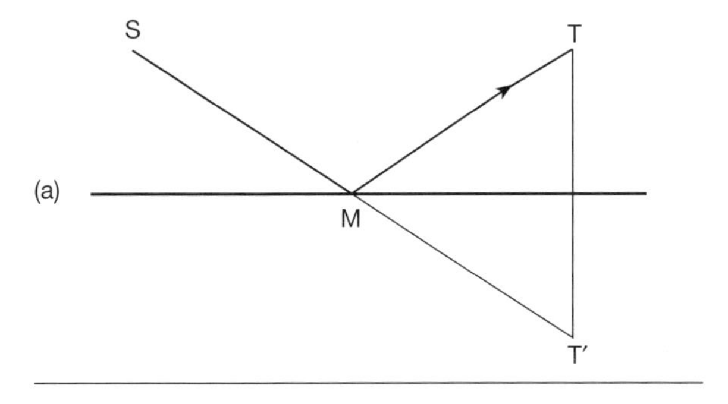

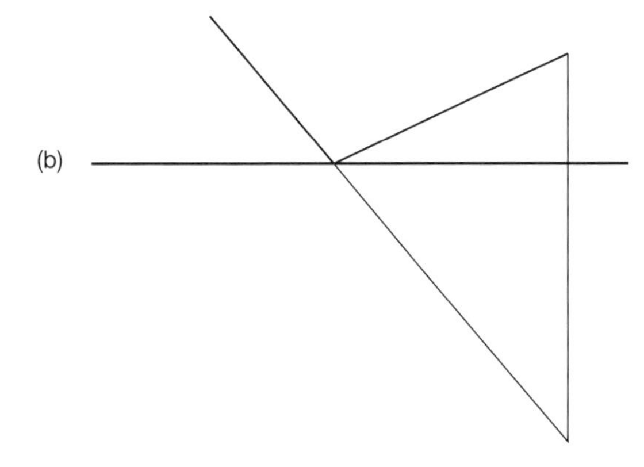

Fig. 1 (a) Light from S takes the shortest path to T via M, where the angle of incidence equals the angle of reflection. This path equals the straight line ST′. (b) Any other position of M makes SMT longer. But how does light 'know' it must strike the mirror at M where the path is shortest?

ties. This duality is the well-known conundrum of quantum physics—namely that rarely occurring photons build up diffraction patterns which 'should' only happen when they interact with each other. Why individual photons show such wave properties seems to be beyond our imagination to grasp, though QED provides rules which describe all the main phenomena of optics with complete accuracy.

When there are just a few photons detected by counters, the probabilities of where a photon will go are represented in QED as arrows in Feynman diagrams. The probabilities of events are given by the squares of the lengths of the arrows, called the 'probability amplitude'. The final path is described with the 'final arrow', given by linking head-to-tail the small arrows representing the possible paths the photon could have taken. The direction is given by an imaginary stop-watch hand, rotating very rapidly while the photon travels (twice as fast for blue as for red light), which stops when the photon reaches the detector. The position of the stopped hand points in the direction the photon takes.

The ultimate mechanism is left mysterious; but what happens is described with a *procedure:* (Feynman, 1985: 37):

> *Draw an arrow for each way [the event could happen], and then combine the arrows ('add' them) by hooking the head of one to the tail of the next. A 'final arrow' is then drawn from the tail of the first arrow to the head of the last one. The final arrow is the one whose square gives the probability of the entire event.*

The direction of the final arrow is the position of the imaginary stop-watch hand timing the photon.

The final arrow is longest when the little arrows point in nearly the same direction, which is when they take the same time. This is the least-time-path—which includes the middle of the mirror—which is where the angle of incidence equals the angle of reflection. So, most light is reflected from where the angle of incidence equals the angle of reflection as this is where neighbouring arrows point in much the same directions, and so add. The reason for the law of reflection is in the nature of the photons of light, which includes a time principle. The arrows of Feynman diagrams are not supposed to exist (as normal objects exist) but they are useful aids to computing what will happen, so are given a significant status.

This probability model is far removed from the Greek's bouncing ball idea (though no doubt *this* has its problems!) and is very different from familiar experience, or common sense.

We will look now at far less subtle aspects of mirrors. But to find appropriate ways of thinking about everyday reflections is surprisingly difficult, though unfamiliar concepts are not needed. We will start with mirror images, leading to the (for most people) puzzling reversals.

Virtual images

Plane mirrors give no image without a focusing eye, or a camera. This was appreciated by Newton in *Opticks* (1704). Newton realized, also, that the virtual image of plane (and convex) mirrors has a psychological component. The virtual image depends on *lack of understanding* by the brain's *visual system*—though we understand *conceptually* what is happening. For we *know* that the reflected object is in front of the mirror though we *see* the image behind it. We see objects as through the mirror though we know they are in front, because the brain's visual processing assumes that light always comes directly from objects—though this is false when the light is bent by reflection or refraction (Fig. 3).

Paradoxically, one experiences two mirror selves: oneself that one can touch in front of the glass, and an insubstantial ghost of oneself behind it.

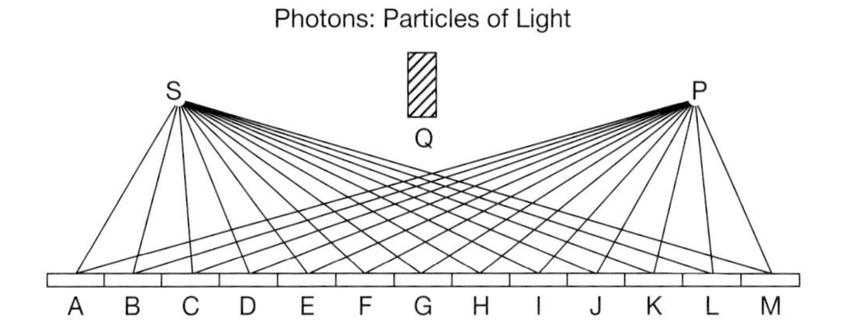

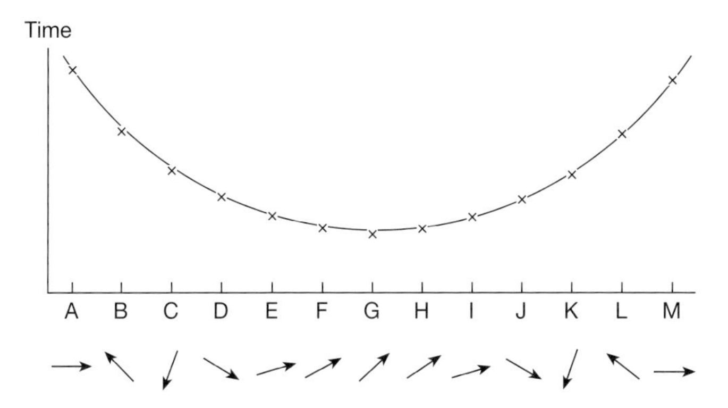

Fig. 2 QED supposes that light reflects from all parts of the mirror. But all except from near the middle cancel out. From Feynman (1985).

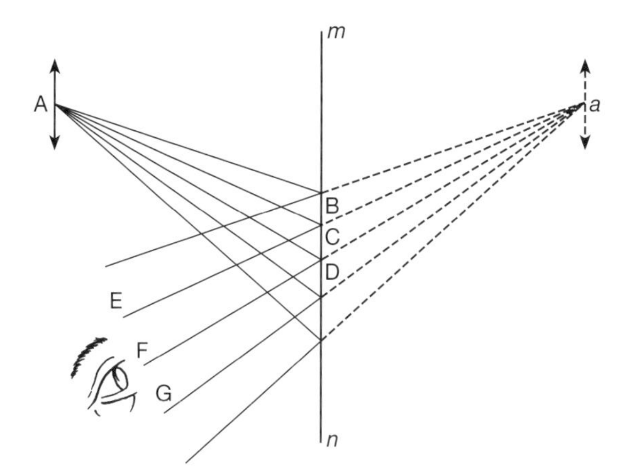

Fig. 3 Newton's drawing showing how a virtual image is formed, *Opticks* (1704). The visual brain assumes that objects lie along the lines of sight. So, though we know the object is in front we see it behind, through the mirror.

It seems very likely that mirror images suggested body—mind dualism: that mind is separated from, and can be independent of the brain. This idea is associated with René Descartes (1596–1650) but is really far older, as it is embalmed in perhaps all religions. Mirrors seem to have had a profound, though deeply misleading, effect on how we see and think of ourselves. Perhaps because of mirrors, it has taken many millennia to accept that mind is brain-based. For most of us now this remains an astonishing, even a shocking, hypothesis—that we carry machinery creating mind in our heads.

Mirror-reversals

Especially confusing, for almost everyone, is the sideways reversal of objects in a looking-glass. Presumably it was just too obvious for Newton as he does not discuss it; yet why symmetrical plane mirrors reverse right–left, though not up–down, has confused practically everyone since Plato and he got it wrong (Plato, *c.* 340 BC). That there is a problem is sometimes denied. But, for example, writing looks very different in a mirror. The reflected E looks reversed (Ǝ), while an M looks the same, seen directly or in the mirror (M). So there is a phenomenon to consider. The puzzle, for most people (once they think about it), is how a symmetrical mirror normal to the observer can produce an asymmetrical reversal. The principle is curiously hard to grasp. People can understand it yet forget the point a week later.

This asymmetry of mirror-reversal seems to violate a basic principle: Curie's principle, that *systematic asymmetry cannot be produced from symmetry*. Either Curie's principle is wrong, or there is some asymmetry in the situation. If the reason is not apparent this everyday experience really should be explained.

The philosopher Immanuel Kant (1724–1804) thought that mirror-reversal is a problem too hard for the human mind to grasp (Kant, 1783). There have been, and there still are, a plethora of obviously false theories passionately defended. Explanations have called upon many sciences: *physics* (of space); *anatomy* (horizontal separation of our eyes, or the near-lateral symmetry of the human body); *physiology* (neural reversal, in eye or brain); *psychology* (mental rotation, in perception or imagination); *linguistics* (ambiguity of the words 'right' and 'left' (advocated by philosophers such as Jonathan Bennett (1970) and critically explored by Block 1974). Block considers all sorts of cases, such as a mirror in a tunnel through the earth; but although what is meant by 'horizontal' or 'vertical' becomes ambiguous in such situations—or when observers are separated around the earth—this does not affect what is going on. It simply makes it harder to describe. There is a revealing difference here between philosophy and science.

Of course, *optics* is often invoked, especially crossings of light paths in the eyes; but these are at all orientations, and so are both horizontal and vertical, and cannot favour either. There are no crossings in a plane mirror. Perhaps surprisingly, ray diagrams cannot help; for a diagram can be oriented in any direction, and so cannot distinguish horizontal from vertical. This is the same for a map. A map can be oriented in any direction, so it cannot tell where north, or any other direction is, without a compass.

In his well-known book *The Ambidextrous Universe* (1965) Martin Gardner introduces mental rotation, suggesting that we rotate ourselves mentally this way because it is easiest, as our bodies are more nearly symmetrical horizontally than vertically. But is this really a psychological phenomenon? Mental rotation as measured (Shepard and Metzler, 1971) is far too slow and inaccurate, and only works for simple objects (Fig. 4).

We see mirror-reversal even when there is no information to alert the brain that a mirror is involved; as when there is no frame, or in a photograph of a reflection. Indeed, the fact that photographs show the reversal precludes any such mental rotation as the cause. Of course, the camera's optics is essentially the same as the eye's; neither can reverse only one axis optically. For various reasons we can rule out all these kinds of accounts (Gregory, 1997).

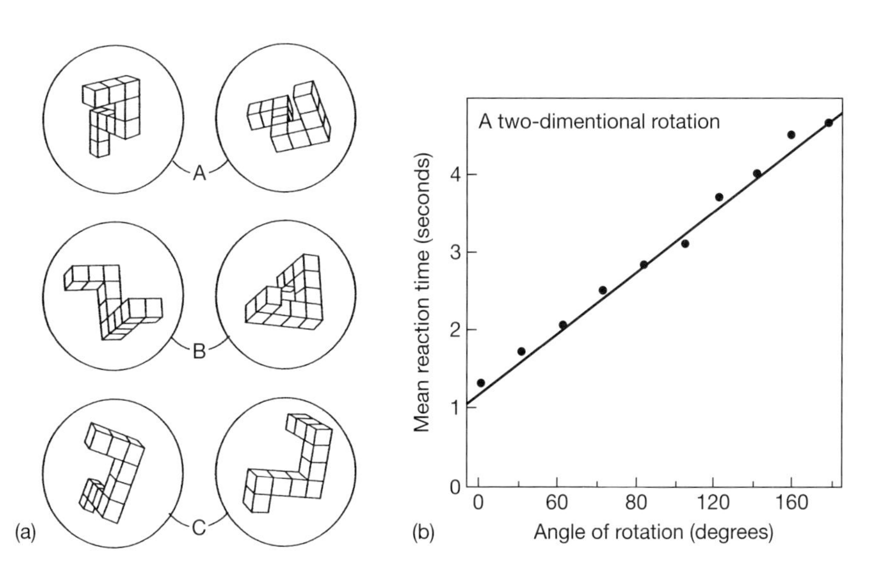

Fig. 4 Mental rotation. Are the two shapes the same object? The experimental subject mentally rotates the first, to judge whether it fits the second. This is slow, the time linearly increasing with the angle—far too slow and unreliable to explain 'mirror reversal'. (From Shepard and Metzler, 1971.)

Reversal in depth

Lateral (and up–down) reversal is very different from the other kind of mirror-reversal—of depth. This is optical, and a ray diagram does give the answer. When you move your hand towards the glass, it recedes away from you as seen directly; but in the looking-glass, it comes towards you. This reversal in depth is due to the optical path-length increasing with object-to-observer distance, but reversed as the object is seen behind the mirror, although it is in front.

Depth-reversal has sometimes been invoked to explain the lateral reversal (Gardner, 1965). But depth-reversal is orthogonal, symmetrical, to both up–down and to right–left, and cannot favour either. So depth-reversal cannot explain the right–left reversal.

To sum up: lateral reversal cannot be explained by light crossing at the mirror—as it doesn't—nor from depth-reversal. For this is symmetrically orthogonal to right–left and up–down, and cannot favour either. Nor from linguistics, or mental rotation. Contrary to some text books of optics, a ray diagram is useless—as it can be oriented in any direction and so cannot tell what is horizontal or vertical. So what is going on? Or, is the answer simply too obvious, to interest us?

The answer

Mirrors allow *backs* of objects to be seen. A particularly clear example of what is happening, is writing on the back of an opaque card seen in a mirror. The mirror must be behind the card (or a book) for the writing at its back to be visible in the mirror. The reversal is caused by the card being *rotated*, from direct view, for the writing to be seen in the mirror behind it. So we see the writing on the back of the card made visible in the mirror, but reversed by the rotation of the object to face the glass. We are apt to forget that the object is rotated, yet this is the simple key to the phenomenon.

Why is the writing *right–left* reversed and not *upside down*? Small objects (such as books) are generally rotated around their vertical axis, as because of gravity this is the easiest rotation. This gives right–left reversal (see Fig. 5). But an object *can* be rotated around its horizontal axis. Then the writing, or whatever, will appear in the mirror upside-down—as it now is—and *not* right–left reversed. What we see depends on how the object is rotated from direct view to face the mirror behind it.

It is important to note that writing on a transparent sheet need not be reversed, because it does not have to be turned around to be visible in the mirror. So it appears the same on the transparent sheet and in the mirror, with no 'mirror-reversal'. This pinpoints the key importance of rotation, producing lateral or up–down reversal.

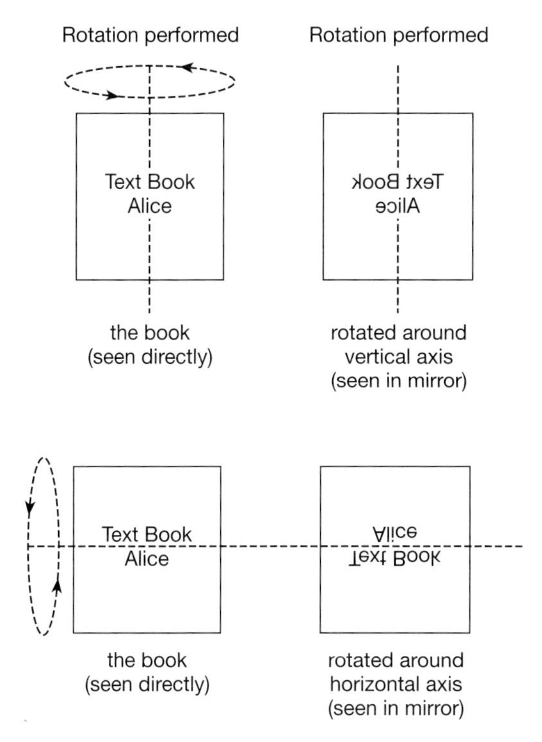

Fig. 5 Rotations producing reversals. The writing makes it easy to recognize the reverals—right–left for vertical-axis and upside-down for horizontal-axis rotations. (From Gregory, 1996.)

The essential principle is: *the reversal of the image is caused by rotation of the object*, from direct view, to face the mirror. Though simple, this is not a familiar kind of explanation. It is not within optics, psychology, linguistics, or indeed any of the mirror-reversal hypotheses drawn from familiar sciences. So explaining the image-reversal by object rotation may not seem attractive. Conversely, familiar and trusted kinds of explanations are seductively misleading, when they draw us away from what turns out to be the answer.

Reversed scenes

It is all very well to say that a book is reversed by being turned to face the mirror—but what of whole rooms, or landscapes? Surely they are not rotated? The most familiar example of scene-reversal is in the driving mirror of a car. The number plates of cars behind appear right–left reversed. But why? What is rotated here?

The car's mirror allows us to see thing behind us while we are looking forward. To see behind directly, you have to turn your head around. When you turn your head to see the cars behind, directly without the mirror, there is no reversal. Then their number plates look normal—not reversed. But while looking in the mirror, your head and eyes are rotated 180° from the direct view behind. The rotation here is of the head and *eyes*, not of objects or the scene.

So object and scene-reversals are both due to rotations—by rotation of objects, or of the observer. The principle is the same though what is rotated is different.

Principles of lateral and depth-reversal

We arrive, however, at two essentially different principles for two essentially different kinds of mirror-reversals: *lateral* (or up–down), and *depth*-reversal. Lateral reversal is given by *rotation* around the vertical axis of objects, or the observer; but depth-reversal, very differently, is given by increased optical path length with increased distance—seen as coming towards one through the mirror, though it is approaching it from in front.

The lateral reversal may be up–down rather than right–left. It is simply false to say that mirrors *always* reverse horizontally. The reversal is, however, always *either* right–left *or* up–down, but never both—for a rotation is only possible around one axis at a time. Right–left reversal is most common, as because of gravity the rotation is usually around the vertical axis. This is easiest for small objects, and we walk around objects on the horizontal ground, and turn our heads around their vertical axis when we are upright. The rotations giving corresponding reversals are shown in Fig. 5.

Up to now, the issues are I believe clear-cut. Indeed, this makes it remarkable that there is so much controversy. We come, however, to a tricky example which I think is not fully understood—seeing one's own face in a mirror. Here we have to speculate on just what is happening, and experiments are needed to clarify several points.

Face-to-face

One's own face is a uniquely strange mirror object, for it is never seen directly, as it is invisible without a mirror. How, then, do you know that it is yourself in the mirror? Experiments by Gordon Gallup (1970) show that children younger than ten months do not recognize themselves in a mirror, and no animals except chimpanzees can do so. Gallup's experimental technique, is to place a spot of rouge on one side of the face and note whether the baby, or animal, touches its own face or the mirror. Babies, and animals except chimps, touch the mirror, not the face. Older babies and human adults, as well as chimps, touch the spot on the face. So they seem to know it is themselves in the mirror.

How does one recognize one's own face, though it is never (normally) seen except in a mirror? It may be that the related movements of the image to one's own movements is the key to initial self-identification. Would young children recognize themselves in a photograph? Would time-delayed video allow self-recognition? I doubt it. These would be interesting experiments.

To see yourself in a mirror, you have to turn around to face it—normally rotating *vertically* as your feet are on the ground.—so you become right–left reversed from how others see you face-on. But how does one see *oneself* 'mirror reversed'— reversed from what? For as one's own face is invisible without a mirror, there is no direct view for comparison. It is this that makes one's face a uniquely puzzling mirror-object.

To appreciate just what is happening, we need a clear criterion for what is or is not reversed. The clearest example is the appearance of writing. Provided we ignore symmetrical letters (especially O and X), writing is immediately recognized as normal, up–down, or right–left reversed. Taking, for example, the word MIRRORS (noting especially the order of the letters and the orientations of M, R, and S) we immediately recognize the possible reversals given by a rotation (as in Fig. 5).

Let us use writing for checking what happens to your face in a looking-glass. A little experiment may help. Ask someone to place writing (such as MIRRORS) on your forehead. The writing will look normal from in front, to the experimenter. To yourself, looking in the mirror, it will appear right–left reversed. The letters and your face are reversed in the mirror by comparison with the experimenter's view looking directly from in front. This is because you are rotated 180° from the experimenter's direction of viewing. (So their left eye is opposite your right eye.)

It might be useful to experiment further. Try writing MIRRORS on a transparent sheet, and ask the experimenter to place this on your forehead, so it looks normal for him or her looking at it directly from in front. Then lift it off, and move it without rotation towards the mirror. This is rather like lifting off your face. Now look at the writing on the transparent sheet, directly (from behind) and also (seen as from in front) reflected in the mirror. Both appear to you right–left reversed. Yet the writing is *not* reversed for the experimenter looking from in front, whose head is rotated 180° from the direction in which you are looking.

You can't see your own face without a mirror because you can't take your eyes out. If you could take your eyes out, and hold them in front of you, they would have to turn around to see your face. Then your eyes would be like the experimenter's looking at you, and it your face would *not* appear reversed.

The looking-glass allows you to see your face without taking your eyes out. The mirror-view is rotated 180° from what would be the direct view of you own face if you took your eyes out, and rotated them to look back at yourself. Your eyes are aimed outward—though they are seeing inward from the mirror in front of you. It is this that gives the right—left reversal of your mirror face from how *others* see you directly. But why, without the give-away writing, do you look reversed to *yourself*—as you do not normally see yourself except in a mirror?

We may question how far this is true. Do you appear reversed to yourself, with no writing? I am not so sure that this is true. Let's try the Gallup technique. Which side of the face in the mirror do you touch? With a spot on my right cheek, I find that I touch the corresponding side of my mirror-image. But this is the *left* cheek of my double. So, am I really seeing my double as myself? Compare this with writing. One soon finds the corresponding letters, and points to them in the mirror, crossing over sideways. This seems to be different from how one sees one's own face.

It is intriguing to compare this with looking in a *non-optically, reversing mirror*, This can be simply a pair of plane mirrors, forming a vertical right-angle corner (Fig. 6b). This must, surely, be useful for artists painting their self-portraits. Rembrandt had to re-paint a self-portrait, as he emerged left-handed!

These variations should be interesting. Although the 90° corner mirror counters the lateral reversal of a looking-glass, it does not counter the depth-inversion. So depth remains reversed. But depth can be *removed*, by using a video camera and TV screen. Rotating the camera 180° to face a mirror behind it will reverse laterally, with no depth. Depth-inversion in the looking-glass, or in the corner mirror, can be *un-reversed* with a mirror placed behind the head, facing the mirror (or mirrors) in front. These arrangements allow all combinations of reversals to be investigated. They invite experiments. Which makes it easier, or even harder, than usual, to knot a bow-tie?

In the optically reversing mirror (Fig. 6b), one continues to point at one's image without cross-over. The reader might like to check this. If this is the same for optically reversing and for optically non-reversing mirrors, do we see ourselves reversed in a looking-glass? I am not sure that we do—at any rate, reliably. For as are only sees one's face in a mirror these is no reference for comparison.

This is different for skilled movements seen in a mirror, such as combing the hair, or shaving, or making up. These are surprisingly difficult in the *non*-reversing mirror. The *absence* of reversal is confusing! This is only so for skills habitually undertaken with a looking-glass. Evidently we learn specific mirror skills. To see how powerful this is, try

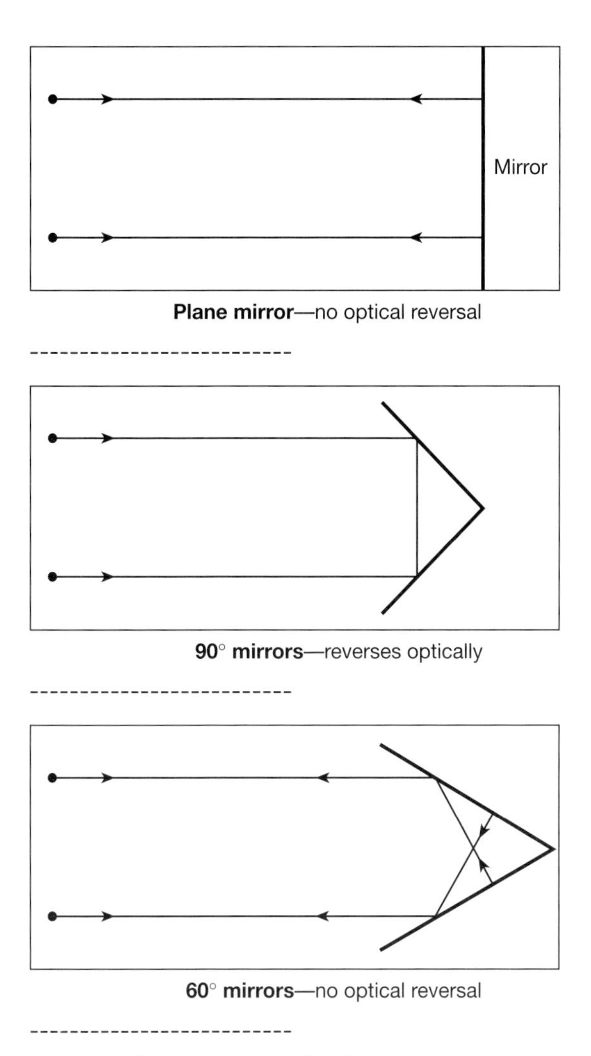

Fig. 6 Optical mirror reversals. The looking-glass has no crossings. The 90° vertical corner mirror reverses right–left (countering looking-glass 'mirror reversal'). The 60° corner mirror is like a looking-glass. (Light from the first mirror meets the second mirror at right angles, and returns to the source.)

tracing a simple design, such as a St David's star, or writing your own name while looking in a mirror. It is incredibly difficult, though familiar shaving and so on are easy.

Although normally you only *see* your face in a mirror, you do *know* your face by touch. Can this serve as a (perhaps not very reliable) reference? In fact there are two touch faces: your *private* inner-face, that you experience by pursing your lips or screwing up your eyes—involving commands to move these muscles—and your to-you invisible *public* outer-face, that you can touch with your fingers. Do we compare the mirror-reflection with our touch faces? If so, which? This is not really understood.

The inner and outer touch faces are another version of the inner—outer ambiguity deep in mirror-reversals. It invites an amusing experiment. With your eyes shut, ask someone (the experimenter) to draw letters lightly on your forehead, so you can feel them. Try to 'see' them on your forehead. When drawn as normal for the experimenter; do you 'see' them as normal, or as reversed? In other words, are they experienced as from without, or from within your head? The result is: either may be experienced. There are individual preferences for 'seeing' as from inside or outside, but it may 'flip'. (Does this locate one's self? Can the sense of self move out of the body?) Evidently seeing oneself is not a simple matter and we need further experiments.

Finally, as depth is reversed, why doesn't your face look *hollow—like* the inside of a mask? For isolated objects (such as one's hands) are reversed in depth in the mirror. Here, again, there is a seductive hypothesis which though true is not relevant. For there is the interesting fact that a hollow mask does not look hollow! At least, from a metre or more away, with both eyes open, it looks convex though it is concave—hollow. But very close to it *does* look (correctly) hollow.

Here we see bottom-up signals from the eyes competing with 'top-down' knowledge, that faces are convex (Gregory 1970). A hollow face is so exceptional, its hollowness is rejected, unless the competing (bottom-up) stereoscopic depth information is very strong.

Look in a mirror, at a hollow mask next to your face. From close up, the mask looks hollow and your face convex. From further away, the mask and your face look convex. The improbability of a hollow face converts the truly concave into an illusory convex mask; though your depth-reversed face is unaffected at any distance from the mirror. Why is this?

For a hollow mask seen directly, the nose is further away than the ears. In the mirror (as for another's face seen directly) your nose looks nearer than your ears. There is no temptation to see yourself as a hollow mask, or as having a hollow face—as you look like someone else facing you—your convex double.

If the reader is confused, experiments with hollow masks, wire objects which reverse in depth and so on should set the matter right—and perhaps reveal more phenomena to explain.

Why are reflections so puzzling?

Perhaps the most puzzling, and indeed important question, is why a familiar looking-glass is so puzzling. This might be because its images are ghosts that cannot be explored by touch. Children have to discover, from exploration by touch and other senses, that objects are hard or hot, or taste good or bad. Then things *look* hard or hot, tasty or disgusting. It is this that allows behaviour to be predictive from vision; for retinal images are themselves mere optical ghosts.[4] Signals from the retinas become imbued with meaning from interactive experiments with objects.

Conversely, we don't learn to see basic object-properties from pictures (or from television, computers, or mirrors) as they deny touch, and so prevent interactive experience. As interaction with objects is so important, we are in danger of losing contact with reality—and so losing meaning—by too many virtual touchless worlds, that cannot be explored. This, in turn, suggests that practical classes in schools and universities, and hands-on science centres for the public, are very important for understanding the world of objects and for learning science. It would be impossible to learn to cook, *only* from television!

A great difficulty for solving problems in any science, is to know where the answer lies before we can see it. Once seen, it may appear obvious and even trivial. Then we can move on to other, perhaps more important questions. The time and effort over centuries spent on puzzling over why a mirror reverses horizontally, shows the importance of finding appropriate ways of thinking. There are all sort of traps and seductive irrelevancies (especially when they are true), which deflect us into tortuous paths leading to dead ends. Another trap is 'ownership'of phonomena, and kinds of answers: here anatomy, psychology, linguistics, optics and several others, have deflected bright people from a simple answer. This may apply to present impossible puzzles, such as consciousness.

If we knew where and how to look, the answer might seem quite simple. The difficulty is to know where to look before an answer is visible. Perhaps we should study, more carefully, signs and symptoms of futile and of promising paths. The best way, as with using a microscope, might be to alternate low-resolution wide fields with detailed narrow views.

Removing mysteries by understanding is progress. As anywhere in science, understanding mirrors does not destroy, but enhances their magic.

Notes

1. This paper is based on my *Mirrors in Mind* (1997) with added reflections, including seeing oneself in a mirror.
2. One can see this effect by shining a light on a polished surface and viewing the refllection with a Polaroid filter which can be rotated. The light dims at a certain orientation of the Polaroid, showing that at a critical angle, reflection polarizes light. This 'Brewster angle' depends on the refractive index of the reflecting material (about 56° for glass). So reflection cannot be simply a surface phenomenon—light 'sees' into the material of the mirror.
3. Light takes the shortest path where it travels most slowly. So it takes a 'short-cut' through high-refractive-index regions, rather than a total straight-line path.
4. They are not, however, seen—as photographs or paintings are seen—as objects; for there are no eyes behind our eyes. This would lead to an infinite progression of eyes looking at pictures in eyes. The retinal image is a cross-section of the visual pathway from world to brain. The retinal image is not an object for perception; so the baby does not need to, and does not, reverse its retinal images, to make the world upright, and the right way round. It relates signals from the eyes to touch—without seeing the pictures in its eyes—though it sees *from* them.

References

Bennett, Jonathan (1970) The difference between right and left, *American Philosophical Quarterly*, **VIII**(3) 175–91.
Block, Ned (1974) Why do mirrors reverse right/left but not up/down?, *Journal of Philosophy*, IXX(1), 259–77.
Feynman, R. (1985) *QED: The Strange Theory of Light and Matter* (Princeton: NJ: Princeton University Press).
Gallup, G.G. Jr (1970) Chimpanzee self-recognition, *Science* 167, 86–7.
Gardner, Martin (1965) *The Ambidextrous Universe* (London: Penguin).
Gregory, R.L. (1966, 5th edition 1997) *Eye and Brain* (Oxford: Oxford University Press).
Gregory, R.L. (1970) *The Intelligent Eye* (London: Weidenfeld & Nicolson).
Gregory, R.L. (1997) *Mirrors in Mind* (Oxford: Oxford University Press/London: Penguin).
Kant, Immanuel (1783) *Prolegomenon to any Future Metaphysics.*
Plato (*c.* 340 BC) *Timaeus*, translated Desmond Lee (1977) (London: Penguin).
Shepard, Roger N. and J. Metzler (1971) Mental rotation of three-dimensional objects, *Cognitive Psychology*, **3**, 701–3.

RICHARD GREGORY

Richard Gregory is Emeritus Professor of Neuropsychology at the University of Bristol. After serving in the RAF (Signals) during the Second World War, he studied Philosophy (Moral Sciences) at

Cambridge, staying on to become a University Lecturer in the Department of Experimental Psychology and Fellow of Corpus Christi College, now Honorary Fellow. He moved to Edinburgh in 1967 to set up the Department of Machine Intelligence, with Donald Michie and Christopher Longuet-Higgins. He then moved to Bristol in 1970 to take up a Personal Chair in Neuropsychology and Directorship of the Brain and Perception Laboratory in the Medical School. His main work has been developing a general account of visual perception, with especial emphasis on a variety of phenomena, including various 'illusions', for suggesting and testing theories of brain mechanisms and cognitive processes. He has invented a hearing aid and a number of optical and recording instruments. Founder of the Exploratory Hands-on-Science Centre in Bristol (now developed with Millennium funding) he was also influential in setting up the Science Centre focused on astronomy at Herstmonceux Castle in East Sussex. He is a CBE and Fellow of the Royal Society. Publications include: *Eye and Brain, The Intelligent Eye, Mind in Science, The Oxford Companion to the Mind* and *Mirrors in Mind*. He is Founder Editor of the journal *Perception*.

How do cells exchange vows, couple, and stay attached to each other?

NANCY J. LANE

1. Historical beginnings

From the moment multicellular systems came into existence, many millions of years ago, it was clearly necessary to forge links between adjacent cells, in order to hold them together as epithelial layers or organs. Plant cells are able to stick together by virtue of their cellulose cell walls, which also serve to provide an external skeletal support. Animal cells lack cell walls and so need to synthesize attachments; without such links, called *intercellular junctions*, animal cells would tend to fall away or part from each other, and this would clearly be deleterious to the integrity of the organism as a whole. This need for cells to 'stay attached' to each other was not really recognized by Schleiden and Schwann when they defined 'the cell theory' back in the mid-1830s, since they were concerned only with the then novel concept of cellular subunits being the building blocks of all multicellular organisms. The resolution of fine detail in biological structure was at that time rather poor in comparison with what can currently be achieved. The light microscope was unable to resolve detail much below 0.2 μm, so that although 'organelles' could be seen to be present in the cytoplasm of cells (as, for example, in preparations made by Golgi in 1898), detailed cellular fine structure could not be properly established. UV microscopes use light of shorter wavelength than normal visible light; this in theory produces enhanced resolution (resolution being inversely proportional to the wavelength of the illuminating source). But it was not until the late twentieth century, with the development of fluorescent probes, bound firmly to specific antibodies, often monoclonal, raised against particular cellular proteins, that it became possible to visualize under the light microscope the precise distribution of the organelles and elements of the cytoskeleton—such as the actin microfilaments or microtubules—in the cellular cytoplasm.

However, with the design and development of the electron microscope (EM) (using electrons, rather than light, as its illuminating source, electrons having a much shorter wavelength than light), and the ancillary techniques to prepare biological material appropriately, which took place from the 1960s onwards, the detailed ultra-structure of cells was finally revealed. These high-powered instruments permitted the resolution of diminutive cytoplasmic structures as separate entities, even when they were only a few ångströms in diameter. As a result, their functional significance could finally begin to be elucidated, since structure and function are intimately correlated.

When animal cells are examined under the EM after they have been fixed in optimal preservative solutions, such as buffered glutaraldehyde, which ensure that all components have been retained, then the cytoskeletal fibres are often found to terminate near certain regions of the external encompassing cell membrane. In epithelial cells, these tend to be the lateral borders of cells that form layers. These concentrations of cytoskeleton near the plasma membrane are part of the cell–cell junctional complexes. They often feature microtubules, seen in cross-section as spheres, and bundles of fibrous microfilaments of actin or other proteins, often only discoverable by biochemical or cytochemical analyses. Intermediate filaments, which are a feature of desmosomal intercellular attachments, tend not to be present in the cells of arthropods, which are the organisms chiefly studied in the investigations considered here.

2. Septate junctions—unique to invertebrates

Throughout the Invertebrata, of which the arthropods are an example of one particularly large and wide-ranging phylum, the cells within organisms are nearly always held together by special zonular or belt-like structures called *septate junctions*; these hold cells firmly in place together but with their membranes at a set distance apart of 18 to 20 nm owing to the characteristic ladder-like striations which straddle the space between adjacent membranes, holding the cells in place and maintaining the regular cleft. These ladder-like striations are actually ribbon-like structures which run between cells and may also act as partial barriers to the free passage of exogenous molecules through this cleft. This cleft is a confined space created by the junctions between the lateral borders of epithelial cell layers, since the cells are held together at their apical ends circumferentially, like a belt. These septate junctional cell contacts, interestingly, are unique to the creatures without backbones, as no vertebrate has ever been found to possess them.

Even the lowly urochordates (tunicates), which are more invertebrate-like in appearance than vertebrate, although they are chordates and hence share ancestry with the vertebrate branch of animal life, do not exhibit these septate junctions. Instead, an alternative form of seal, called the 'occluding' or 'tight' junction, occurs on the lateral borders between the cells of urochordates and of all chordates. Although they are belt-like in distribution too, these are fundamentally different from the septate junctions since they form seals by adjacent cell membrane *fusion* at punctate points of membrane apposition, to prevent the passage of materials between the lateral borders of cells. The adjacent cell membranes are thereby 'sealed' together with no visible intercellular cleft remaining. Septate junctions, on the other hand, would appear to effectively block the movement of substances between cells by means of their ladder-like ribbons that straddle the 18–20 nm space between cellular membranes. These ribbons may be fenestrated, with holes, or may carry a charge, so their capacity to block the translateral migration of materials in the space between cell borders will vary according to the size and/or charge of the molecules attempting entry. The septal ribbons themselves appear to be held in place in the adjacent membranes by being pegged onto transmembrane proteins, which lie in rows within the membrane bilayer, and appear as intramembranous particles (IMPs) when visualized in replicas by freeze-fracture techniques. The intercellular septal ribbons and the intramembranous particle rows are seen to be arrayed in similar alignments, so the former could readily be held in position by the latter. To demonstrate the intercellular cleft in which the ribbons lie, electron-opaque molecules, such as those of lanthanum, may be allowed to infiltrate the space between the cells. This reveals, in a negative-stained fashion, the components of the septal ribbons which stand out in unstained non-opaque relief against the electron-opaque background of the dense tracer in the remainder of the cleft. Models can be constructed from these kinds of microscopical analyses which contain the features both of the ribbon-laden intercellular cleft shown in tracer-infiltrated sections and of the intramembranous features of aligned particle rows, demonstrated in the freeze-fracture replicas.

Biochemical isolation of the septate junctions enables analyses to be made of the molecular weight of their component proteins (Lane and Dilworth, 1989), and by rapid freezing and deep etching, their association with actin on the cytoplasmic membrane surface can be seen (Dallai, Lupetti and Lane 1998). It seems likely that these cytoskeletal actin microfilaments have some involvement in holding the IMPs in place in the membrane, which IMPs in turn hold the ribbons in position in the cleft, since when cells are treated with actin-disruptive agents such as cytochalasin, the septate junctions are seen to become disorganized, as if released from some restraining attachment site (Lane and Flores, 1988).

3. Tight junctions

There are only a few invertebrates that exhibit tight junctions: thus far only certain arthropods with sophisticated central nervous systems (CNS)—those where, it must be assumed, it is particularly important to form effective permeability barriers, such as in the modified glial cells around the brains of spiders, scorpions, and certain insects (Lane, 1992; 2001). The tight or occluding junctions between modified glial cells in these systems are the basis for the blood–brain barrier and occur between the borders of the perineurium glial cells which surround the CNS and thereby protect the nerve cells from any variabilities in the circulating haemolymph. By excluding any extraneous exogenous molecules from the immediate surface of the neurons, there can be no resultant fluctuations in their environment. Hence the neurons, whose activity depends on the existence of a specific resting potential across their membranes, maintained by active ionic pumping, can continue to function effectively. It would seem that septate junctions cannot produce a barrier that is as 'tight' as that formed by the occluding tight junctions. The possession of a physiologically tight barrier would appear to be imperative in the formation of the permeability barriers that vertebrates exhibit in a variety of organs, such as kidney, liver, and gut, since all these possess tight junctions.

So, although some invertebrates exhibit tight junctions, no vertebrates possess the septate junctions which are therefore presumed to be physiologically inferior to the tight junction in establishing efficient barriers.

In those arthropods in which genuine tight junctions can be demonstrated, these junctions have all the same structural features as the vertebrate tight junctions. Hence, in thin sections they possess punctate appositions where the adjacent membranes appear fused. After lanthanum infiltration, the tracer is stopped by these membrane fusions and prevented from further passage along the intercellular cleft. In freeze-fracture replicas, the area of membrane apposition is characterized by a network of moniliform ridges with a complementary set of furrows. This network may sometimes exhibit discontinuities suggesting that the junctions may in some cases be leaky, and this is paralleled by tracer leakage past incomplete membrane fusions. These close membranous contacts are associated with actin bundles which are seen to be inserting into the membranes at the point of punctate contact. Whether these are linked in to membrane-bound occludin and/or claudin, the tight-junctional proteins extracted from the tight junctions of vertebrates by the school of Furuse in Japan (Furuse *et al.*, 1993; Tsukita and Furuse, 1999), is not clear. Studies are currently under way to establish if antibodies to these normally vertebrate tight-junctional proteins are specifically localized in the invertebrate tight-junctional membranous appositions as well.

4. Junctional/IMP migration

Patterns of IMP distribution in the membrane faces of cells can be followed when stem cells are dividing into new epithelial cells which subsequently establish position and contact between existing cells. In the case of septate junctions, their putative association with half-ribbons, held in place in adjacent cell membranes, would allow each cell membrane to contribute one half-ribbon to the maturing septate-junctional ribbon structures. The IMPs which are in the membrane, acting, it is thought, to hold the ribbons in position, take up a series of arrangements over developmental time which are transformed from single IMPs into short strands of IMPs which gradually take up the mature pattern of the septate junction with its ordered configuration of extensive parallel IMP rows that have been established as being associated with the septal ribbons.

During embryonic development, the arthropod tight-junctional IMPs can also be seen to shift in position by translateral migration (Lane, 1981). In putative junctional areas, they are seen to be transformed from individual particles to ones aligned in short rods or partial ridges, to linear alignments extending for considerable distances. In this way the mature networks of moniliform ridges come to be formed, appearing to be in every way comparable to those of the vertebrates, It is, of course, assumed that, during this process, the adjacent cell membranes are becoming appropriately intimately aligned with one another.

Such changing patterns of IMPs are therefore seen over time with the development of both septate and tight junctions, but the difference between them is that the latter lack septa. The tight-junctional IMPs come to fuse directly with the comparable opposing IMPs in adjacent membranes, whilst the septate-junctional IMPs are linked via the septal ribbons over a cleft of 18 to 20 nm. Although, as already indicated, septate junctions are never found in vertebrates, tight junctions are to be found in certain select invertebrates, so that the latter junctional type must have evolved independently at least twice. The septate junctions occur between all epithelia—the surface integument, the gut, the cells of the salivary gland, ovarian follicles, testes, Malpighian tubules, and so on—of the invertebrates. Although they may exhibit some differences, depending on their embryonic origin, ectodermal versus endodermal, septate junctions are all characterized by the definitive 18–20 nm cleft between adjacent membranes together with septal ribbons traversing the cleft.

5. Gap junctions

But how do cells exchange vows and couple? This is a more profound form of cell–cell interaction in that it involves ionic or metabolic exchange between the cytoplasms of adjacent cells. For this reason the junctions involved are sometimes called 'communicating' junctions.

In cells of similar type, such as occur in a layer of epithelial cells, or in any given organ, adjacent cell membranes are often found in very close apposition, with cell–cell contacts in the form of macular or spot-like configurations, rather than a belt. So close are the cells together, that, although the gap between membranes is not obliterated, as in tight junctions, it is reduced to only 3 to 4 μm. Hence their name, as only a slight 'gap' remains in the cleft. In this area, adjacent membranes are found to possess plaques of clustered transmembranous proteins each of which has a hexameric substructure. In the centre of each hexamer is a tiny pore or channel. These hexamers, called 'connexons' in vertebrate tissues, are positioned so that the pores in abutting connexons from adjacent cell membranes, are aligned. Hence each membrane channel is in continuity with that in the adjacent cell membrane, and ions and molecules are able to pass through the apposed channels. The cells are then said to be 'coupled', since molecules can move from one cell to another, and if these molecules are morphogenetic or signalling in nature, they may exert an effect on the cells to which they are passed. Cells that are coupled can therefore act in an integrated fashion since they will all receive similar messages. Should they cease this exchange for any reason, the coupling is said to be 'turned off', by, it is considered, a configurational change in the six subunits around each central channel. This effectively closes that channel down. This sort of closure can be effected by increases in Ca^{2+} concentration, or decreases in pH, for example. Each of the six hexameric subunits is made up of a protein which crosses the membrane four times; the four portions traversing the membrane are believed to be in the form of α-helices. The configurational change that takes place during channel closure is thought to be by a modification of the position of the innermost α-helix of each hexameric subunit of the proteins which straddle the membrane.

In the tissues of invertebrates, as in vertebrates, gap junctions are common, but although they ostensibly have a hexameric substructure, they seem to be composed of a quite distinct protein, not 'connexin' as in vertebrate connexons, but 'ductin', of lower molecular weight (16–18 kDa) and of a quite different amino acid sequence, thereby exhibiting no homology to the 'connexin' of vertebrates (Finbow *et al.*, 1994). On the other hand, ductin does have sequence homology with a CNS-derived vacuolar ATPase (adenosine triphosphatase), as well as the F_0 (transmembrane) component of ATP synthetase.

When the protein 'ductin' is sequenced, hydrophobicity plots suggest, as mentioned above, that it straddles the membrane four times, each time with an α-helical configuration (Finbow *et al.*, 1994). It is assumed that this is the composition of each of the six subunits or hexamers, that make up a single gap-junctional particle, many of which are then aggregated in macular plaques, as seen in freeze-fracture replicas.

In replicas, these particles can be seen to be extremely numerous and clustered in a number of plaques of quite different dimensions. In other words, different numbers of gap-junction particles make up every plaque, each of which is assumed to be a functional gap junction, with each component particle containing a central channel for the exchange of molecules. However, in static replicas, we cannot judge if coupling is actually occurring, so many of these channels could be resting or non-functional. There do seem to be rather more particles than one might have predicted were required for cell coupling. But why are some plaques made up of tens, and others of hundreds, of particles? These questions cannot yet be answered.

Gap junctions can, however, be visualized in the process of formation, by particle aggregation. In embryonic tissues, mature gap junctions are not to be found. As the embryo's organs develop, gap junctions begin to appear, but initially by the appearance of single gap-junctional particles, in presumptive junctional areas, presumably by the insertion of an individual gap-junction IMP into the lipid bilayer membrane. At later stages, the gap-junctional IMPs appear to coalesce and form loose clusters. With time these aggregate into irregular particle plaques and only with a further period of time do they come to take up the regular rounded configuration that is typical of mature gap junctions. One startling observation is that they seem to have no cytoskeletal attachments, unlike the membranes of septate and tight junctions which link onto actin filaments, so that there is no obvious mechanism in gap junctions to pull the IMPs into plaque position! There must be an attraction of some sort between the junctional particles which leads them to aggregate preferentially together.

Concluding remarks

These various junctions all co-exist on the same membrane face. Astonishingly, they seem never to make an error in the process of aggregation. Although their component particles are all of a rather similar size in freeze-fracture (8–10 nm per IMP for tight junctions and septate junctions, 13 nm per IMP for gap junctions) they clearly are composed of very different proteins, even though we do not yet know the exact chemical details of the septate-junction transmembrane protein. Moreover, the junctions exhibit a flexibility in that they can disaggregate and reassemble. This is particularly striking in the insect transition from the larval to the pupal stage when both gap and tight junctions are studied in the CNS (Lane and Swales, 1978a, b). The larval junctions break down and exhibit particle disaggregation as they enter the pupal stage. When the larval CNS is transformed, during pupation, into the very different adult CNS, the junctional particles re-aggregate into functional mature structures. It seems likely that the cytoskeleton is very important in the regulation of junctional assembly—this is certainly the case in that demonstrable actin microfilament links can be seen running into the septate junctions and the tight junctions. In the latter it seems likely that the actin's action is mediated by other proteins to which it is bound in 'piggy-back' fashion, such as cingulin, occludin, claudin, and others (Isukita and Furuse, 1999).

It is clear that intercellular junctions are important to cellular well-being, given their ubiquity throughout the animal kingdom. The differences we observe between the vertebrate and the invertebrate junctions seems not to be in the 'format' of the junction complexes (except for the septate junctions) but in the amino acid sequence of the proteins. It would seem that 'coupling' and 'sealing', as well as adhering, are crucial to cellular health. What is surprising is the rigorous limitation whereby septate junctions are restricted entirely to the invertebrates. Whether they were a primitive form of seal that was later usurped by the vertebrate tight junctions, with the tight junctions of sophisticated arthropods developing as a comparable structure only over time, cannot now be definitively established. Work from the laboratory of Peter Bryant in California shows that some *Drosophila* genes have an impact on the emergence of normal septate-junctional structure since in mutants, such as *Discs large* tumour suppressor protein, they are improperly formed; they are also homologous to parts of the tight-junctional ZO-l protein sequence, which encodes a guanylate kinase homologue, suggesting they both may be involved in cell signalling (Woods and Bryant, 1991; Willott *et al.*, 1993). Further studies on the existence of the vertebrate junctional proteins in invertebrate tissues, by immunocytochemistry and gene sequencing, are required to be able to fully understand these junctional relationships.

References

Dallai, R.D., Lupetti, P., and Lane, N.J. 1998. The organization of actin in the apical region of insect midgut cells after deep-etching. *J. Structural Biology*, 122, 283–92.

Finbow, M.E., Goodwin, S., Meagher, L., Lane N.J., Keen, J., Findlay, J.B.C., and Kaiser, K. 1994. Evidence that the 16 kDa proteolipid (Subunit C) of the vacuolar H+-ATPase and ductin from gap junctions are the same polypeptide in *Drosophila* and *Manduca*: molecular cloning of the *Vha16k* gene from *Drosophila. J. Cell Sci.*, 107(7), 1817–24.

Furuse, M., Hirase, T., Itoh, M., Nagafuchi, A., Yonemura, S., *et al.* 1993. Occludin: a novel integral membrane protein localizing at tight junctions. *J. Cell Biol.*, 123(6), 1777–88.

Lane, N.J. 1981. Evidence for two separate categories of junctional particle during the concurrent formation of tight and gap junctions. *J. Ult. Res.*, 77, 54–65.

Lane, N.J. 1992. Anatomy of the tight junction: invertebrates. Chapter 3 in *Tight Junctions* (ed. M. Cereijido) CRC Press, Ann Arbor, pp. 23–48

Lane, N. J. 2001. As above. Chapter 3, (Ed. Anderson and Cereijido) CRC Press Ann Arbor, pp 37–79.

Lane N.J. and Dilworth, S. 1989. Isolation and biochemical characterization of septate junctions. Differences between the proteins in smooth and pleated varieties. *J. Cell Sci.*, 93, 123–31.

Lane, N.J. and Flores, V. 1988. Actin filaments are associated with the septate junctions of invertebrates. *Tissue and Cell*, 20(2), 211–17.

Lane, N.J. and Swales, L.S. 1978a. Changes in the blood-brain barrier of the central nervous system in the blowfly during development, with special reference to the formation and disaggregation of gap and tight junctions. I. Larval development. *Develop. Biol.*, 62, 389–414.

Lane, N.J. and Swales, L.S. 1978b. Changes in the blood-brain barrier of the blowfly during development, with special reference to the formation and disaggregation of gap and tight functions. II. Pupal development and adult flies. *Develop. Biol.*, 62, 415–31.

Tsukita, S., and Furuse, M. 1999. Occludin and claudins in tight junction strands: leading or supporting players? *Trends in Cell Biol.*, 9, 268–73.

Willott, E., Balda, M.S., Fanning, A.S., Jameson, B., VanItallie, C., and Anderson, J.M. 1993. The tight junction protein Z0-l is homologous to the *Drosophila* discs large tumour supressor protein of septate junctions. *Proc. Natl. Acad. Sci. USA*, 90, 7834–8.

Woods, D.F. and Bryant, P.J. 1991. The discs-large tumor suppressor gene of *Drosophila* encodes a guanylate kinase homolog localized at septate junctions. *Cell*, 66, 451–64.

NANCY J. LANE

Dr Nancy J. Lane is a cell biologist at Cambridge University and an Official Fellow and Lecturer at Girton College. She is also a Fellow of the Institute of Biology, the Zoological Society of London (ZSL) and the RSA. She has published widely and has edited *Cell Biology International*. The Prime Minister appointed Dr Lane to the Citizens Charter Advisory panel and she also chaired the Working Party on Women in Science, Engineering and Technology for the Cabinet Office. She now serves as Deputy–Chair of University UK's *Athena Project* and is President-elect of the Institute of Biology. She was awarded the OBE in 1994 for services to science.

Fluorine, the ultimate combiner

JOHN H. HOLLOWAY

The elements at the extreme ends of the periodic table are generally the most exceptional and have the most extreme properties. Although Humphrey Davy, then aged 29, succeeded in isolating metallic potassium and sodium on the extreme left of the table in 1807 and, on the other side of the table, provided proof of the elementary nature of chlorine in 1810, his concerted efforts to prise fluorine out of its compounds by both chemical and electrolytic methods failed. The work was carried out in the laboratories of the Royal Institution and it is clear from his account that his health must have suffered as a result of these efforts:

> *I applied the power of the great voltaic batteries of the Royal Institution to the liquid fluoric acid ... The manner in which the surrounding atmosphere became filled with the fumes of the fluoric acid, rendered it, indeed, very difficult to examine the results of any of these experiments ... and I suffered considerable inconvenience from their effects during this investigation. By mere exposure to them in their uncondensed state, my fingers became sore beneath the nails, and they produced a most painful sensation which lasted for some hours, when they came into contact with my eyes.*[1]

Worse befell others. In the 1830s, the Irish brothers, Thomas and George J. Knox, suffered horrifically as a result of inhaling HF vapour, George losing his voice permanently. The deaths of the Belgian, P. Louyet (1850), and of F. J. Nicklès of Nancy (1869) appear to be as a direct result of breathing the vapour. Despite these appalling events, B. Brauner of Prague endeavoured to test for fluorine by inhaling the gas liberated from the heating of the acid salt K_3HPbF_8 and noting whether 'fumes of hydrofluoric acid issued from the nose'.[2]

Some 70 years after Davy's experiments, on Saturday, 26th June 1886, the French chemist, Henri Moissan, finally devised an experiment whereby the fluorine he obtained was not lost by reaction with its surroundings before it was detected and examined. He carried out the electrolysis of anhydrous hydrofluoric acid in a U-shaped platinum tube at −50°C. He characterized hydrogen liberated at the negative electrode and showed that the gas produced at the positive electrode gave a yellow compound with mercury, decomposed water with the production of ozone, spontaneously ignited phosphorus, sulphur and silicon, and liberated chlorine from KCl.[3] So it was that the ninth element in the periodic table was finally isolated and the first of its extraordinary properties identified. Notable above all, and clearly demonstrated in Moissan's first experiments, is its phenomenal capacity to form compounds. This is neatly encapsulated in Vernon Gibson's elegant quatrain:[4]

Mistress Fluorine

Fervid Fluorine, though just Nine,
Knows her aim in life: combine!
In fact, of things that like to mingle,
None's less likely to stay single.

Now, more than one hundred years after the isolation of the element, chemists have learned to respect fluorine but they have also learned to manipulate it. It cannot be worked with comfortably 'on the bench', but it can be manipulated in a vacuum line made of nickel, some nickel alloys, stainless steel, copper or brass (see Fig. 1), all of which form a protective fluoride layer, or in plastics such as PTFE (polytetrafluoroethylene) or Kel-F® (trifluorochloroethylene) which themselves contain fluorine and are stable to attack by it.

Working with fluorine in an enclosed all-metal system is not without difficulties! Not least is the fact that one cannot see what is going on inside the reactor—one cannot watch reactions taking place. However, I can show you that what one might see if you could peer into the confines of the reaction chamber is very exciting indeed (see Fig. 2).

Part of the joy of being a chemist is that one tries to understand why particular reactions occur in the way they do. Why is it that fluorine reacts so violently with so many other elements and compounds? Why do fires and explosions result when one brings fluorine into contact with so many other materials? The first reason is that the fluorine–fluorine bond is very weak. The heat of dissociation of fluorine is only 153 kJ mol^{-1}. This means that it readily produces highly reactive fluorine atoms or radicals. On the other hand, the bonds it forms to other elements are very strong. When carbon burns in fluorine to form CF_4, the average bond energy of the C–F bonds formed is 490 kJ mol^{-1}. Similarly, the average bond dissociation energies for the S–F bond in SF_6, for the I–F bond in IF_5 or IF_7 and the H–F bond in HF are all very high (see

Fig. 1. Metal vacuum line for manipulating volatile fluorides.

Fig. 2. Carbon and sulphur burning in fluorine.

Table 1). Indeed, the strength of the single bond in HF is similar in energy to that of the triple bond in nitrogen. This means that there is a significant thermodynamic driving force underlying most reactions involving fluorine. It also means that HF, CF_4, SF_6, and materials like Teflon®, $(CF_2–CF_2)_n$, which contain strong element-to-fluorine bonds, are thermodynamically very stable.

One of the reasons for the strength of element-to-fluorine bonds is that fluorine is the most electronegative element known. The extraordinary tendency it has to attract electrons to itself means that many fluorides are ionic and those that are covalently bonded tend to have a degree of ionic

$$D(F-F) = 153 \text{ kJ mol}^{-1}$$

c.f.

$$D(S-F) = 330 \text{ kJ mol}^{-1}$$

$$D(C-F) = 490 \text{ kJ mol}^{-1}$$

$$D(I-F) = 264 \text{ kJ mol}^{-1}$$

$$D(H-F) = 570 \text{ kJ mol}^{-1}$$

Table 1. The heat of dissociation of fluorine and the average heats of formation of element-to-fluorine bonds for SF_6, CF_4, IF_5 and IF_7, and HF.

character within the covalent bond, which provides an additional component to the overall bond energy.

Another characteristic of fluorine is its small size. It is bigger than hydrogen but smaller than all the other elements in the first row of the periodic table. In fact, it has only about half the radius of the first element in that row (see Table 2). Fluorine is also much smaller than the other halogens; for example, the covalent fluorine atom is only about one quarter of the size of chlorine and a tenth the size of iodine. This means that fluorine can reach places that most other elements cannot reach! This not only enhances its ability to react, but also means that its chemistry differs from that of the other halogens. For example, it forms compounds that have no parallels in chlorine, bromine or iodine chemistry. Thus SF_6 is known but SCl_6 and SBr_6 are not, IF_7, ReF_7, and OsF_7 have

PERIOD 1	Li	Be	B	C	N	O	F
	1.34	0.90	0.82	0.77	0.75	0.73	0.72
PERIOD 2	Na	Mg	Al	Si	P	S	Cl
	1.54	1.30	1.25	1.17	1.10	1.02	0.99
PERIOD 3	K	Ca	Ga	Ge	As	Se	Br
	1.96	1.74	1.26	1.22	1.21	1.16	1.14

Table 2. Size trends in periods 1, 2, and 3 (covalent radii in Å).

been prepared but their chlorine equivalents do not exist. And so it is that fluorine embodies in the one element a number of extraordinary properties (see Table 3) that all contribute to make it highly reactive. It is not surprising, therefore, that it was fluorine that held the key to the preparation, in 1962, of noble-gas compounds. While working on the preparation of PtF_6 by fluorinating platinum metal in glass apparatus, Neil Bartlett noticed another compound which at first he thought was an oxide fluoride of platinum. Eventually, he showed that it was, in fact, a novel species, $[O_2]^+[PtF_6]^-$. He also demonstrated that it could be prepared by the simple expedient of mixing PtF_6 vapour with oxygen. This compound was remarkable in two ways. The first was that the compound contained oxidized oxygen, $[O_2]^+$, a species that many regarded as highly unlikely (since oxygen, surely, should not itself be easily oxidized!). The second was that PtF_6 is evidently an exceptionally powerful oxidizing agent. Bartlett knew, of course, that in making $[O_2]^+$ by removing one electron from neutral O_2, the energy involved was, by definition, the first ionization potential of oxygen. Therefore, to identify other species that might be similarly oxidized by PtF_6, he merely had to look up a table of first ionization potentials. When he did this, he was surprised to find that the element xenon, a so-called 'inert gas' had almost the same first ionization potential as oxygen, but he carried out the crucial experiment of mixing gaseous xenon with PtF_6 vapour. The fact that a brown *solid* was immediately precipitated all over the inner walls of the vessel was evidence enough that a xenon compound had been produced. Others soon confirmed the reaction and also carried out a related reaction with a new and even more powerfully oxidizing hexafluoride, RuF_6. Amongst the products, green crystals were observed which were indicative of the formation of RuF_5. This suggested that some fluorine may have been taken up in another side-reaction with the xenon and that there was a real possibility that a simple xenon fluoride might be prepared. Henry Selig, the late John Malm, and Howard Claassen took a volume of xenon, mixed it with five volumes of fluorine under pressure and heated it to 400 °C. Fortuitously, they had selected the perfect stoicheiometry and conditions to produce XeF_4.[5]

- The heat of dissociation of F_2 is small
- The heat formation of bonds of fluorine is often large
- Fluorine is the most electronegative element
- The F atom is small
- The F^- ion is small

Table 3. The extraordinary properties of fluorine.

As fluorine chemistry has developed, it has become apparent that the compounds produced by the reactions of fluorine with other elements or compounds are of two types. They are either, as outlined above, inert, stable species or, alternatively, they may behave rather like fluorine itself in that they are capable of fluorinating other elements or compounds. The reasons for this will become evident later!

The second joy of being a chemist is the challenge of making use of the products of research. Despite the extraordinary difficulties in handling fluorine and its compounds, many fluorine species have become important and influential in many areas of our lives. We are all aware of the benefits of the incorporation of small amounts of fluorine in teeth. Its introduction into the water supplies of some regions of the country as fluorosilicic acid, H_2SiF_6, or sodium fluoride, NaF, is controversial, but its incorporation as NaF or sodium fluorophosphate, Na_2PO_3F, into fluoride toothpaste is generally seen as beneficial. Such toothpaste is widely used. Similarly, most people know that Teflon, famed for its non-stick properties and first introduced as a non-stick surface in frying pans, is a fluorine compound. However, fluorine compounds are much more widely used than is generally appreciated.

Let us begin with some of the simple compounds we met at the beginning. Sulphur hexafluoride, SF_6, we learned, is a molecule with strong S–F bonds. It is an extraordinary compound because, as might be expected, the strong attachments of the fluorines to the central sulphur render it stable and unreactive. Because it is stable and unreactive it does not interact with our nasal detectors and so it is odourless, if breathed it can suffocate us but it is not toxic, and because it will not react further with most things it is, therefore, non-inflammable. Amazingly, despite its mass, this very stable compound is also a gas. This is because the molecule is perfectly octahedral in shape (see Fig. 3) and is, therefore, without an in-built dipole which might result in its coalescing with other SF_6 molecules to produce a liquid or a solid. Also, because the small fluorines distributed so symmetrically about the sulphur are resistant to any induced dipole, the SF_6 molecule behaves like a very hard ball. This means that if two SF_6 molecules are brought-together they tend to bounce off one another. Such behaviour is the fundamental characteristic of gaseous molecules. Since, as well as being gaseous and inert, SF_6 also has high dielectric strength, it finds a use as an insulating gas in high-voltage circuit-breakers, protecting the terminals from corrosion through aerial oxidation by the flash when the switch is thrown.

When fluorine and sulphur interact it is also possible to produce SF_4. This, unlike its sister, is a mild fluorinating agent. The reasons are that, in this case, the four fluorines do not completely surround the central sulphur (see Fig. 3), so that the electrons of this central atom can interact with the electrons of other atoms, and there is also potential for the

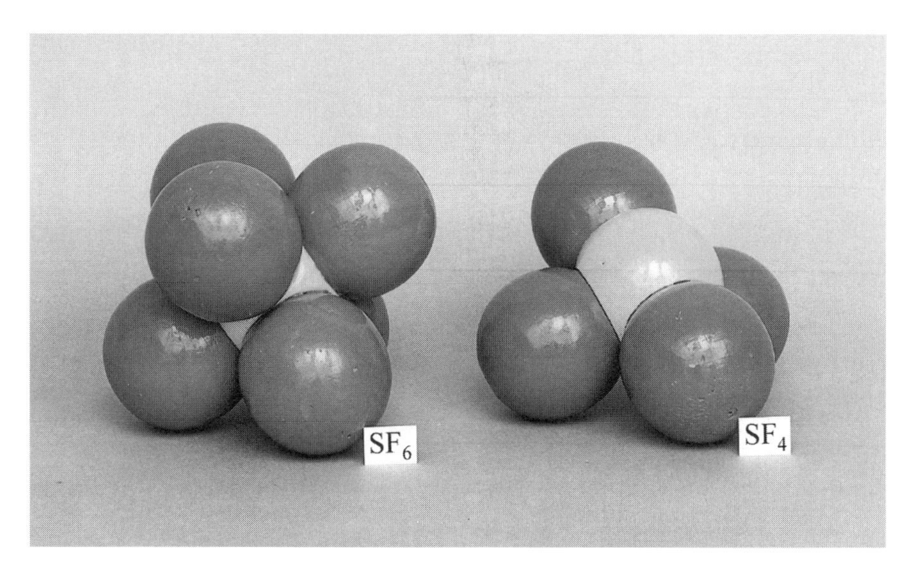

Fig. 3 Models of SF_6 and SF_4.

central sulphur to make attachments with up to two more other elements. So SF_4 is used to introduce small amounts of fluorine into the surfaces of natural, neoprene, acrylonitrile-butadiene and styrene-butadiene rubbers to give high-lubricity, Teflon-like surfaces suitable for windscreen wiper blades or non-stick gaskets. It can be similarly employed to fluorinate the surfaces of cellulose products like wood, paper, cotton, wool, polyvinyl alcohol, and cellophane, where it replaces some OH groups with the non-polarizable F atom to provide materials with increased water repellency.

Another compound we met at the beginning was CF_4 with its strong C–F bonds. Polytetrafluoroethylene, or Teflon $[(CF_2\text{–}CF_2)_2]$, is a chain compound consisting of rows of carbons to which F atoms are attached, while CF_4 is a single carbon to which F atoms are attached. In each case the C atom is attached to four other atoms which are either fluorines or carbons. Because the carbon chain of Teflon is surrounded by small non-polarizable fluorine atoms, the surface tends to 'repel' other materials and to be slippery. Hence, because it is both stable and slippery, it is used industrially as a gasket material, as a surface in many kinds of cookware, for surfacing the bottom of skis and, by virtue of its ability to throw off water and dirt, as a surface finish in modern structures such as the Millennium Dome.

Short-chain-length versions of this polymer are the basis of products such as Scotchgard and Milease which, when applied to carpets, render them easier to clean up spillages from and to remove stains. As aerosol sprays, these same materials are used to make clothing, shoes and boots waterproof. A cleverly engineered form of Teflon known as Gore-Tex, is the basis for what is arguably the best all-weather clothing used by mountaineers and hikers, sailors and pot-holers, not to mention its use in modern army combat uniforms, astronauts' suits and the best-quality surgical gowns (see Fig. 4).

The molecular-sized holes in the fibrillar structure of Gore-Tex (see Fig. 5) has also allowed widespread application in the production of material for use in surgery, the size of the holes being arranged such that tissue can grow into the non-toxic, highly stable, inert material. Thus, in humans and other animals Gore-Tex can be employed as heart patches and for vascular grafts (see Fig. 6).

The infamous CFCs, chlorofluorocarbons, such as CF_2Cl_2 and $CFCl_3$, are also fluorine-containing compounds which are directly and closely

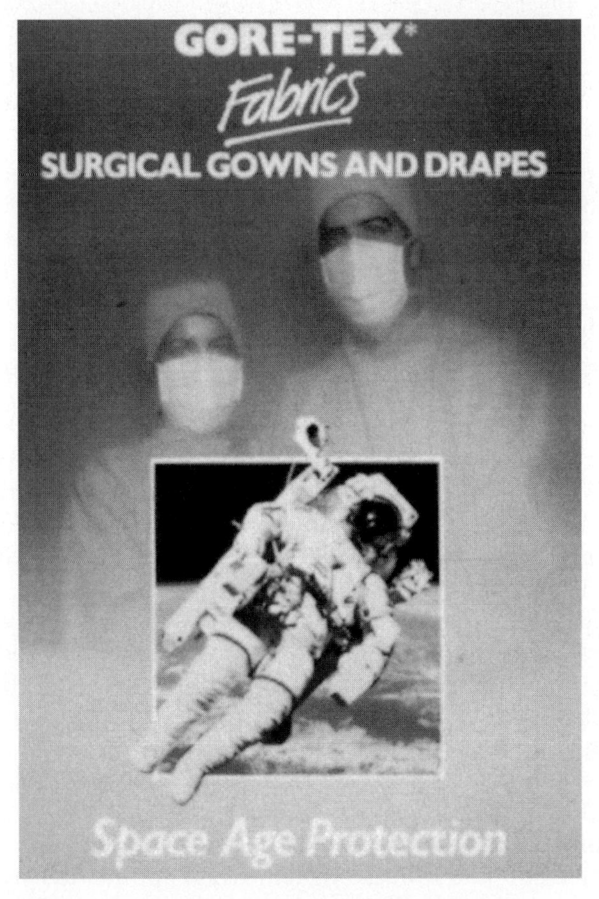

Fig. 4 An advertisement for Gore-Tex surgical gowns.

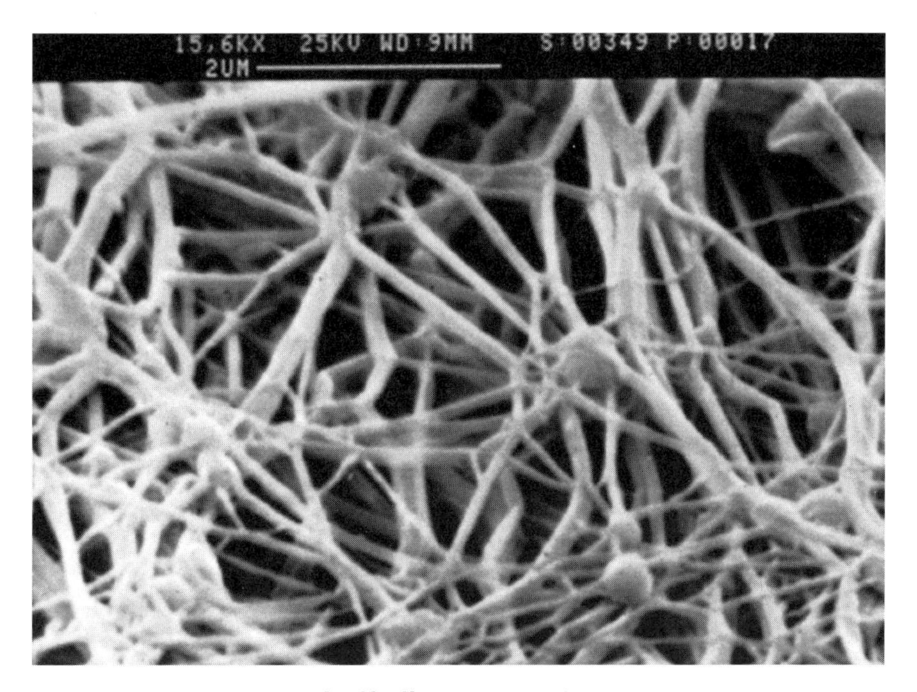

Fig. 5 The fibrillar structure of Gore-Tex.

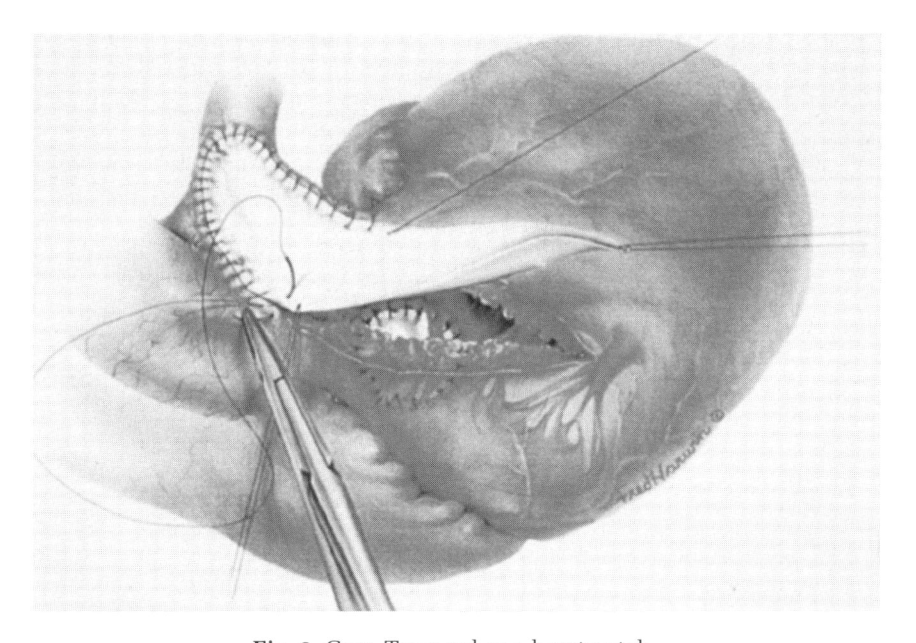

Fig. 6 Gore-Tex used as a heart patch.

related to CF_4 (that is, in one case two fluorines and in the other case three fluorines are replaced by chlorines) and have similar properties. Like SF_6, these molecules are gases. They were first created as non-toxic, non-flammable refrigerants to replace SO_2 (sulphur dioxide) and NH_3 (ammonia), the toxic and highly corrosive materials that were originally used. Later, because of their apparent harmlessness and ease of containment they were used as aerosol propellants. The convenience and low cost of these materials in *both* applications endeared them to people and huge quantities were used. In the UK alone around 200 million cans of aerosol propellants were being consumed each year. These same compounds also found widespread industrial applications as foam-blowing agents, degreasing agents and solvents. A related compound, bromochlorodifluoromethane, CF_2ClBr, (a bromchlorofluorocarbon) is perhaps the best fire-fighting agent for petrol or other solvent fires that has ever been invented and is still in use for fighting aircraft-engine fires. Concerns about damage that these compounds might cause to the ozone layer came first from modelling of possible chemical reactions that might be occurring in the upper atmosphere. Before any real proof was obtained such was the projected seriousness of the situation that governments united under the Montreal Protocol to ban or limit the use of CFCs and other chemical relatives.

Eventually, evidence of chlorine monoxide, ClO, was found in the upper atmosphere. This suggested that the problem lay in the weakness of the C–Cl bonds in the chlorofluorocarbons (see Fig. 7). Now, CFC replacements, in which as many as possible of the chlorine atoms are replaced by hydrogen or fluorine (which, as we have learned, forms a much stronger bond to carbon) are being used. These are not damaging or are less damaging to the ozone layer.

One of the most important contributions of chemistry to the quality of human life has been the creation of compounds that induce anaesthesia. Modern inhalation anaesthetics are almost entirely dependent on fluorine chemistry. The first was Fluothane, $CF_3CHClBr$, again a relative

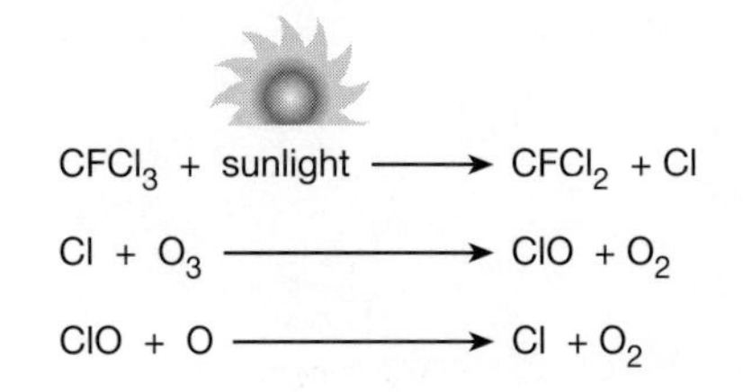

Fig. 7 The reaction of $CFCl_3$ induced by sunlight in the upper atmosphere, showing the breaking of the C–Cl bond and how this results in the destruction of ozone.

of the CFCs and an ozone depleter, but it was used in 70–80% of all anaesthesias for a long time and has been employed in an estimated 500 million surgical operations since 1956. In recent years it has been replaced by a new class of fluorine-containing chemicals called 'fluoro-ethers' (see Fig. 8) which are not only not damaging to the ozone layer but also have none of the side effects such as the post-operative nausea experienced by many who were treated with Fluothane.

Organic fluorine chemistry is already widely applied in medicine and biochemistry. The fluorine atom mimics the hydrogen atom with respect to its space requirements at enzyme receptor sites, its strong tendency to attract electrons to itself can change the chemical behaviour of other reaction centres in molecules, and the substitution of hydrogen attached to carbon by fluorine when the carbon is close to a reaction centre may inhibit metabolism because of the high carbon-bond strength. Consequently, fluorine is a component of a wide range of modern drugs including anti-cancer and anti-viral agents, anti-inflammatory drugs, antibiotics, central nervous system agents, diuretics and antihypertensive agents, and antiarrhythmic heart drugs. Because the substitution of fluorine for hydrogen can increase lipid solubility and so enhance the rates of absorption and transport of drugs, some fluorine-containing drugs are more effective than their hydrogen-containing counterparts. Fluorine also finds wide application in agriculture in the plant protection field.

In the field of medicine, fluorine chemistry holds a number of promises for the future. Fluorocarbon liquid emulsions containing perfluorodecalin, $C_{10}F_{18}$, or perfluotri-*n*-propylamine, $(n\text{-}C_3F_7)_3N$, can dissolve and carry oxygen and are already being exploited in assisting the survival of prematurely born babies and offer the prospect of application as artificial blood.

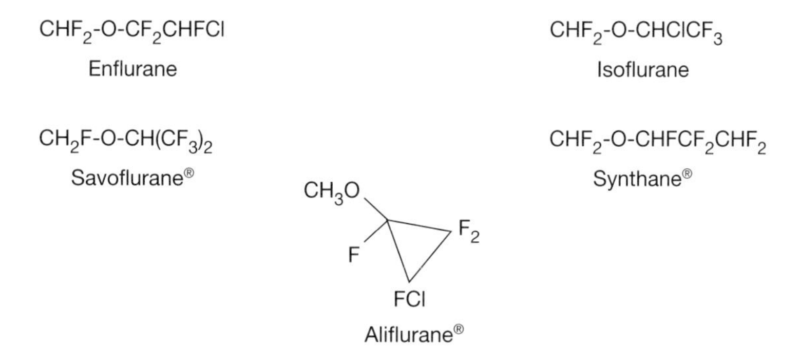

Fig. 8 Inhalation anaesthetics—the new generation of fluoroethers.

The importance of fluorine extends into many other areas: it continues to be crucial in the extraction of aluminium and in the nuclear industry, it has a significant role in the polymer industry, and its applications in modern medicine, pharmaceuticals, and agriculture continue to grow. It also underpins microelectronics and computing where it is found in modern optical fibres, etchants for microchips, and degreasing agents for circuit boards. Currently, work in the University of Leicester and elsewhere[6] promises a new way of carrying out catalytic conversions more efficiently. The principle underpinning the new catalytic system is that the catalyst is designed to be soluble in a fluorinated solvent by attaching to it long carbon–fluorine chains which are sometimes called 'fluorous pony-tails'. The reagents that are to be brought into combination are soluble in an organic solvent which at room temperature is immiscible with the fluorous phase. When, however, the two phases are warmed together they form a single homogeneous phase and reaction can occur under the influence of the catalyst. After reaction is complete, the mixture is allowed to cool and the products, uncontaminated with catalyst, are collected from the organic phase, and the catalyst is isolated in the fluorous phase (Fig. 9). Given that it is estimated that 30% of the GNP for an industrial nation depends on catalysis, this could have far-reaching consequences.

And so fluorine, this most reactive of elements—this ultimate combiner—has been brought under control. Its chemistry has yielded extraordinary and useful compounds, many of which are at the forefront of modern technology in a number of fields.

- Catalyst is anchored in the fluorous phase by the addition of perfluorinated 'ponytails'.
- At elevated temperatures a single phase can be obtained allowing reactions to proceed homogeneously
- On cooling the two phases are regenerated and facile catalyst separation can be achieved

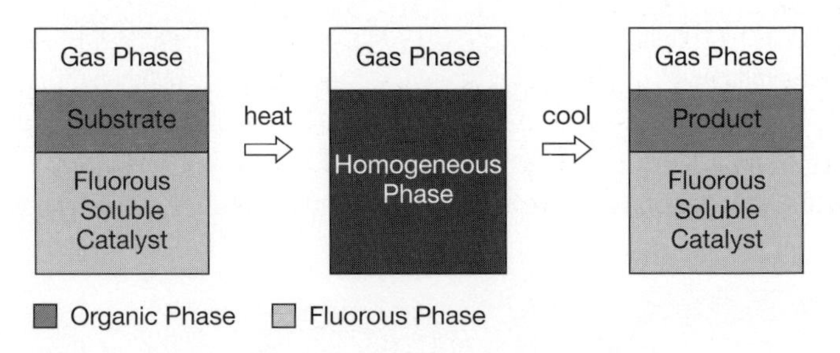

Fig. 9 Fluorous biphase systems (FBS).

References

1. H. Davy, *Phil. Trans.*, 1813, **103**, 272
2. R. E. Banks, Isolation of fluorine by Moissan: setting the scene, Chapter 1 in *Fluorine, The First Hundred Years (1886–1986)*, eds R.E. Banks, D.W.A. Sharp, and J.C. Tatlow, Elsevier Sequoia, Lausanne, 1986
3. J. Flahaut and C. Viel, The life and scientific work of Henri Moissan, Chapter 2 in *Fluorine, The First Hundred Years (l886–1986)*, eds R.E. Banks, D.W.A. Sharp, and J.C. Tatlow, Elsevier Sequoia, Lausanne, 1986
4. The quatrain, 'Mistress Fluorine' is reproduced by permission of Vernon C.J. Newton from his delightful set of verses *Adam's Atoms; Making Light of the Elements*, Viking Press, New York, 1965 (now out of print).
5. J.H. Holloway, *Noble-Gas Chemistry*, Methuen, London, l968.
6. E.G. Hope and A.M. Stuart, Chemistry in perfluorinated organic solvents, Chapter 13 in *Advanced Inorganic Fluorides: Synthesis, Characterization and Applications*, eds T. Nakajima, B. Zemva, and A. Tressaud, Elsevier, Amsterdam, 2000.

JOHN H. HOLLOWAY

Born in 1938 in Ashby-de-la-Zouch, England, John Holloway was educated at the University of Birmingham (BSc 1960; PhD 1963). He was awarded a DSc in 1976. Elected a Fellow of the Royal Society of Chemistry he moved from a lectureship in the University of Aberdeen to the University of Leicester in 1971, becoming Professor of Inorganic Chemistry. He was Head of Department 1987–96. He is an authority on inorganic fluorine chemistry and the chemistry of the noble gases. He has published more than 250 research papers in these areas and one book. He has held the position of Visiting Professor of Chemistry at McMaster University, Canada and The Free University of Berlin and has also been Visiting Scientist at Argonne National Laboratory, USA, the Centre Nucléaires de Saclay, France and the Atomic Weapons Research Establishment, Harwell, UK. He is an Associate of the Jozef Stefan Institute, Ljubljana, Slovenia. He has held public appointments within UK chemical science and education including the Engineering and Physical Science Research Council, the Higher Education Funding Council for England, the Royal Society of Chemistry and the Government's Foresight Panel for Chemistry. Very committed to the popularization of science and chemistry Professor Holloway is a popular lecturer with schools and lay audiences within the UK. He has also contributed to a number of television programmes on chemistry.

Science and society: what scientists and the public can learn from each other

ROBERT M. WORCESTER

Introduction

In *naturenews+*, the daily report of the ICSU/UNESCO World Conference on Science in Budapest on 28 June 1999 published by the *Nature* magazine, the publication's leader made a number of points which this audience might consider: that 'the relationship between the natural and the so-called "social" sciences has never ... been a particularly comfortable one' and that 'some physical scientists ... often find it difficult to accept that another, but equally valid, form of reality can be revealed by what are often disparagingly called the "soft sciences"' ... and that 'they have provided a deep understanding of the dynamics of human society that, in its own way, has made an equally fundamental contribution to our social well-being'.

Last year I joined with some fifty social scientists who went to Budapest to join the some 1600 scientists at the World Conference in an effort to try to bridge the gap. Nobody paid much attention despite the good efforts of *Nature*. Following Budapest, the Foundation for Science and Technology provided me with a platform to bring the findings my colleagues and I had collected to that elite group of scientists, and reprinted my lecture in their Journal, some of which I intend to repeat this evening, with apologies to anyone here who was in attendance at that event or who read them in the Foundation's newsletter. I was also privileged to be invited to give testimony to the House of Lords Sub-Committee under the chairmanship of Lord Jenkin which produced the Science and Society Report in February 2000. Even they acknowledge that '"hard science" is paying increasing attention to insights from social science'.

I intend to develop three themes in this paper:

- methodology of survey research
- trust in science and scientists
- attitudes to animal experimentation

There is a great deal I am not able to fit into this short paper, to do with the public's concern over GMOs, reactions to the BSE crisis, attitudes to the environment among the public, environmental journalists, the City, MPs, and other opinion leaders, public and captains of industries' views about corporate social responsibility, international comparisons, etc. all of which I'd hoped to fit in, but, alas, there's little enough time to cover the ground on these three subjects and interrelate and summarize them, and draw the implications from them I hope to leave you with.

Research methodology

The term 'public opinion' is most often meant to describe the adult population in a one-person–one-vote model, which would have been unheard of a century ago. Yet Abraham Lincoln not only used the concepts of extended franchise and democratic involvement, 'of the people, by the people, and for the people', in the Gettysburg address, he is also quoted as saying, 'Public opinion is everything'.

I am convinced that as a tree falling in the forest makes a noise whether anybody is there to hear it or not, so public opinion exists, perhaps unheard until someone listens. It seems to me that a simple definition will do: 'Public opinion is the view of a [representative sample of a] defined population' (Worcester 1997). The words in brackets delimit the difference between 'public opinion' and a 'public opinion poll'.

There are three things we can measure with the tools of our trade:

- **behaviour**, what people do;
- **knowledge**, what they know, or think they know;
- and their **views**.

I define 'views' at three levels, rather too poetically I fear for scholarly adoption, as

- *opinions*: the ripples on the surface of the public's consciousness—shallow and easily changed;
- *attitudes*: the currents below the surface—deeper and stronger;
- and *values*: the deep tides of public mood, slow to change, but powerful.

Opinions are those low-salience, little-thought-about reactions to pollsters' questions about issues of the day, easily manipulated by question-wording or the news of the day, not very important to the respondent, not vital to their well-being or that of their family, unlikely to have been the topic of discussion or debate between them and their relations, friends and workmates, easily blown about by the winds of the media.

Attitudes, the current below the surface, derive in my view from a deeper level of consciousness, are held with some conviction, and are likely to have been held for some time and after thought or discussion; they are harder to confront or confound.

Values are the deepest of all—the powerful tides of individual and group beliefs, learned parentally in many cases, formed early in life, and not likely to change—which only harden as we grow older. These include belief in God, views about abortion or the death penalty, family values, and the like. It is almost impossible for these to be changed by persuasion, by media discussion or propaganda, or by the positions and arguments of political debate, except over long periods, concerted thought and discussions, a feeling that one is out of step with others one knows and respects, new evidence, changing circumstances or behavioural experiences.

Public opinion research is a simple business really: all you have to do is ask the right questions ... of the right sample ... and add up the figures correctly. My colleagues hate to hear me say that, as their experience, and mine, is that few people, and fewer journalists, commentators or even clients understand how complicated it really is. Also, how the 'marriage of the art of asking questions and the science of sampling' can be misrepresented, especially by the BBC, who will talk about 'stopping people in the street, asking simple yes/no questions'—something they tend to do, but we don't!

When I started MORI, the market research company, here in Britain in 1969, I was puzzled by the apparent unwillingness of some of my clients, mostly commercial in those days, to understand the difference between 'fact' or 'truth' and 'perception', and it was the latter they had to deal with, whether they liked it or not. I was constantly having to remind my clients, in every boardroom presentation, that I was not measuring some sort of abstract 'truth', and that it did them no good to say 'that's just not so', when confronted by unpleasant or unpalatable findings from surveys of their customers or their employees. Coming back depressed from just such a presentation one time early on, one of my colleagues said, 'Not to worry, if you don't like the findings, it's because you either asked the wrong sample or asked the wrong questions!'. Reflecting on that, I started prefacing nearly every presentation with the reminder that I did not offer them 'truth', but 'perceptions', and reminded them that it was the perceptions of their customers, their workforce, their other publics of importance to them, that that was what they had to deal with, and that I could bring two findings to them: one, where something was believed to be so, and negative, if it was in fact wrong, it was a misconception or misunderstanding that could be dealt with by communications, but if objectively the negative finding was correct, then it was the reality that had to be changed, or the organization had to live with the consequences.

It wasn't until some years later that I stumbled over the wise words of the first-century slave-philosopher, Eptitus: 'Perceptions are truth, because people believe them'.

Quantitative v. qualitative

Of recent notoriety is the much-mocked 'focus group', of political frame. The Jenkin Report correctly states, 'As a polling method, focus groups are of limited value'. True. It is one of several techniques known as *qualitative* research, derived from psychology, principally the depth interview, one on one, and the focus group. Quantitative research on the other hand is structures, systematic, and replicable, using defined sampling techniques and various questionnaire constructions, and is derived from sociology. Briefly, quantitative research tells you 'who', and 'what', and 'when', and 'where'; qualitative research helps you to understand 'why'. There is some overlap, but 'that's' what each is best used for. The practitioners who get into trouble are those who do not observe the methodological boundaries, and use the wrong technique for the wrong reasons, e.g. the journalists who describe the People's Panel we do for the Cabinet Office as a '5000-strong focus group', or the political consultants who use group discussions with a handful of 'floating voters' and declaim 'public opinion is moving against (or towards) the war/Euro/policy or issue of the moment'.

One uses qualitative research to provide an insight into issues under examination, to dig below the surface to discover the reasons: Why? or Why not? One tries to understand the impact of communications and to obtain feedback from the public most relevant to the enquiry, not necessarily by any means the 'general public', but perhaps 'activists' in political research, not the committed, but the 'floating voter', who determines the outcome both in party and in number of seats in every general election, or OAPs, or lost customers, or key employees, or decision makers.

Qualitative research, especially the focus group, can enable the moderator, or for that matter, and preferably, other members of the group, to challenge initial reactions and comments, and these techniques enable us to collect relevant, and sometimes surprising, verbatim comments. It must be remembered, however, that the results of focus groups and depth interviews are inevitably anecdotal, and are not representative of public opinion generally, and cannot be replicated with a known degree of statistical certainty.

While individual depth interviews and focus groups typically are face-to-face, they may be done on the telephone, and have been known to be done postally in an iterative manner, and experiments are now being undertaken using the Internet. They are normally semi-structured, the moderator using a topic guide which includes both direct and indirect questioning techniques, and projective techniques as well, to provide insight and 'peel the layers of the onion' of people's thinking. There are many modifications to the standard qualitative study, including mini-groups, extended and even repetitive designs, placement of stimulus and call-back, pre–post experiments, paired depths, and increasingly such variations as citizens' juries, citizens' workshops, and the like.

What qualitative research is not able to do is provide any statistical numerical measurement, allow robust subgroup comparisons, provide accurate measurement of changes over time, or compare individual responses. To do such things, a quantitative approach must be undertaken.

Quantitative studies allow within the limits of sampling, short of a 100% census, for statistically representative and reliable, and replicable, hard data which can be used to provide subgroup comparisons, track aggregate change over time, or, with a panel methodology, even track individual change over time. These studies can be face-to-face, conducted over the telephone, by post, by self-completion questionnaire, and increasingly, on the Internet or otherwise using computer technology.

As I said earlier, survey research is the marriage of the art of asking questions and the science of sampling. First, looking at questionnaire design, a good question should:

- Be relevant to the respondent
- Be easily understood by the respondent
- Be unambiguous in meaning
- Mean the same to the client, researcher, and *all* respondents
- Relate to the survey objectives

And, above all, it must be as unbiased as the English language and human fallibility will allow. Not, for instance, to fall into the trap that the founder of Gallup here in Britain, the late Henry Durrant, did in the first year of the British Institute of Public Opinion (BIPO), later to be known as the Gallup Poll, in 1938, when he asked the British public 'Are you in favour of direct retaliatory action against Franco's piracy?', and in just eleven words broke five rules of what we would know now as good questionnaire construction. We ask today, 'Are you in favour, or do you oppose?', thus balancing the options; we wouldn't use such an ill-defined phrase as'direct retaliatory action', as one person's might be a punch on the nose, and another is a bullet between the eyes; and what percentage of the British public would have known the meaning of the word 'retaliatory' in 1928, anyway? The fourth rule is to define who's who. We still would today say 'Tony Blair, the prime minister', or 'the former prime minister', when referring to Lady Thatcher or John Major, and in 1938 certainly would have reminded the respondents who Franco was. And fifth, 'piracy'? Rather pejorative, and we do eschew pejoratives these days!

There are many question types: multiple-choice, involving showcards in face-to-face interviewing, selecting from a list, preferably fewer than a dozen items, questions of degree, classifying people on one side of the fence or the other, and measuring the strength of feeling on issues, revealing how strongly they feel, which if done properly ensures balanced responses and is good for tracking change over time, and may be analysed by using the political concept of 'swing', of which more later.

Another type of question is the ubiquitous agree/disagree, or Lickert, scale, which posits a contentious statement, and asks the respondent to agree or disagree with it, preferably asking them whether they agree strongly, or tend to agree, or disagree, for that matter. Used correctly, balanced, and adjectively modified, it is one of the best tools we have for tracking change over time. There are many things you can do with survey research, but getting people to recall what they thought six months, much less six years ago isn't one of them. Tests have shown that what they think they thought then is likely what they think now, not what they thought!

Then there are spontaneous questions, obtaining their own words to open-ended questions either writing down verbatim what is said, or using interviewer pre-codes, which help the interviewer and aids analysis, is faster, so is cost-effective, and removes one step in the analysis process.

What we don't do is ask 'simple yes/no' questions, except on matters of behaviour, such as 'Have you driven a car in the past seven days?'.

Sampling is the more scientific part of this marriage, working from the laws of probability, and I'm sure that there are in this audience more expert statisticians than I would claim to be. This audience will have been trained in the science of statistics, but that is not true of the radio or television journalist or the majority of their audience. The analogy I use most is the bottle of wine, saying that whether it is a 'split' or a 'jeroboam', it only takes one sip out of the bottle to tell whether the wine is sweet or sour. I then go on to say, if I get two sentences before they cut me off, that with a population of roughly 3 million adults in Ireland, 40 million in Britain, or nearly 200 million in America, the statistical reliability of a sample of 1000 people has the same sampling tolerance in each country, if the sample is drawn properly.

There are, as you will know, varying techniques of sampling, random (not haphazard, as most people think is the meaning of the word, while we take it to mean equal probability of selection), quota samples, censuses, snowballing, visitors, etc. Much of our work other than for the government is quota sampling: quicker, cheaper, easier, and, in a fast-moving election, more accurate than the traditional alternative, probability sampling, which, with all the money and time in the world, is to be preferred, but who ever has either? For the most part, our commercial and media surveys use quota-controlled samples using enumeration districts as the basic sampling frame, thus probability at the first stage, and within enumeration district, controlling for gender, age, social class defined by the occupation of the respondent, home ownership, sometimes car(s) in household, telephone, work status, etc.

Sample size is determined usually not by the total sample reliability desired, from top down, but from the bottom up, determining which subgroups need examination with reasonable reliability. It was never more true the generalization that I've taken to heart out of John Maddox's latest book, *What Remains to Be Discovered*, that 'Scientists faced with the need to estimate quantities that are prone to error are familiar with the problem: no measurement can be absolutely accurate!'.

As a frame of reference, and things really aren't that simple; a sample size of about 2000 people is accurate to plus or minus two per cent 95 times in a hundred, one of 1000 to plus or minus three per cent 95 times in a hundred, and a sample size of 500 will be accurate to plus or minus five per cent 95 times in a hundred. Or, as I say to the British public, most of whom will wager from time to time, 'I'll give you 19 to 1 that our poll for the *Times* (of say 2000 adults) will be accurate to plus or minus three per cent, a 2 to 1 bet it will be within two per cent, and evens that it will be within one per cent.' Not precise, but it carries the degree of confidence closely enough.

There are a number of ways that we can judge and analyse the data, beyond seeing the 'top-line' results:

- Internal comparisons (demographic, geographic, attitudinal)
- Normative data (if available)
- Longitudinal studies (over time)
- Ideal versus actual (within questionnaire)
- Anticipated outcome (guestimate)
- Ideal outcome (realistic)
- International comparisons (if relevant)

We start the analysis process by first looking at the 'top-line' results, based on the sample as a whole, and then look at the subgroup findings, and then can carry out any number of multivariate techniques or other sophisticated analyses which help identify patterns, themes, trends perhaps, or other additional insights into the meaning of what people have told us.

'Science and Society'

In their report, 'Science and Society' (2000), the House of Lords' Select Committee on Science and Technology quoted at length from survey data on British attitudes to science and scientists, much of which does not require replication here. Their conclusion was that

> *Society's relationship with science is in a critical phase ... public confidence in scientific advice to Government has been rocked by BSE; and many people are uneasy about the rapid advance of areas such as biotechnology and IT ... This crisis of confidence is of great importance both to British society and to British science.*

I agree, from my reading of the data to which I have access, and the data which has been released to the public conducted by others, although getting hold of such data is not as easy as it might, and should, be. Too often, government agencies are slow to release data (to be fair, this is not true of either the Cabinet Office or OST), or to release it at all, and sometimes when it is released, it is months, even years, after its utility has expired except for historic purposes. As the report put it:

> *The administrative culture of the United Kingdom is notoriously secretive ... there is an abiding presumption that the government information and decision-making processes are confidential and closed. This has left the field wide open to allegations of conspiracy and cover-up.*

The report, which I call the Jenkin Report after the chairman of the Sub-Committee which took testimony, Lord (Patrick) Jenkin, makes a number of valid points. I commend a careful reading of the Jenkin Report, and to a careful thought about the implications of their recommendations, and a careful thought too for its limitations, for as I have critiqued it elsewhere, it says little about the role in science and society of the private sector, which funds more research than does the public sector, or the role of the civic sector, which certainly makes more noise about it than does either the government or private sector.

Among their other conclusions:

- Public interest in science is high
- People now question all authority, including scientific authority (*sic*)
- People place more trust in science which is seen as 'independent'
- There is still a culture of governmental and institutional secrecy in the UK, which invites suspicion
- Some issues are treated too narrowly as 'scientific', when non-scientific considerations are insufficiently considered
- Risk and control are misunderstood by all sides
- Underlying people's attitudes are people's values
- There has been a cultural change in the attitude of most British scientists, in favour of public outreach activities (unproven in my view)
- Suppressing uncertainty is bound to diminish public trust and respect
- Transparency is important

- The UK must change existing institutional terms of reference and procedures to open them up to more substantial influence and effective inputs from diverse groups
- Scientists must learn to work with the media as they are
- The culture of UK science needs a sea change in favour of open and positive communication with the media.

The Committee has made some 26 recommendations. It is but a beginning.

In my testimony to the Committee, I argued five theses.

1. **The British public tends to judge the value of scientific advances by their end purpose**. If no end purpose is made clear to them, many tend to implicitly assume that it has no useful purpose or even that its purpose will be detrimental rather than beneficial. The intensity of ethical objections to particular work, for example the use of animals in experimentation, is similarly significantly affected by understanding of what it is hoped will be achieved and alternatives to causing animal suffering in aid of scientific advancement.
2. **Scientific developments aimed directly at achieving improvements in human health-care are the most valued by the public**. However, the public is often ill informed about the purpose of scientific experimentation, and public opinion is less supportive than it otherwise might be because not enough people instinctively make the connection between means and ends. Research for its own sake, and particularly research seen primarily as having a commercial motive, is unpopular.
3. **Ignorance about the way in which science is regulated and restricted leads many of the public to assume that the regulation is insufficient, and this in turn makes them more likely to be hostile to science**. Yet they are eager to receive such information and show intelligent interest when they do so. Regulatory bodies whose work was well publicized and which were seen to be free of control by government or other vested interests might significantly improve the climate of public opinion.
4. **There is scepticism and mistrust in government and business alike, and although a majority of the public say they trust 'scientists' whenever a scientist's employer or sponsor is mentioned, the veracity of the source becomes highly relevant: the scientists trusted by the highest proportion of people are those working for environmental NGOs**. It is clear that many of the public assume (perhaps not consciously) that scientists cannot maintain their independence, integrity or objectivity when working for an interested party. Furthermore, in most fields of public controversy, the government is regarded as an interested party, and neither it nor scientists seen to be working for it are trusted by a majority of the public.

5. **Significant numbers of the public are prepared to use their power as consumers to put pressure on those involved when they object to a scientific procedure or principle.** Science is important to people, and they understand that it is. We are all affected by science, from today's weather to global warming, from developing world famine to GMOs, from new developments in medical research to space exploration, but we know what we don't know, and suspect those that do. That is human nature, and scientists must understand that in the world of the twenty-first century, it is no longer acceptable to have the good of mankind at heart, and to be seen to have the good of mankind at heart. If scientists do not do so, they run the risk of public scepticism at best, cynicism somewhere in the middle, and distrust, suspicion, and negative reaction at worst.

I will skip lightly over much of the evidence from which I draw these conclusions, to concentrate only on two themes, trust in science and scientists, and the public's views on scientific experimentation on live animals.

Thesis 1: The British public tends to judge the value of scientific advances by their end purpose

What do the public think of as 'science'? The first factor to be considered in understanding the public's attitudes to any subject is how they 'instinctively' interpret references to the subject itself. In the case of science, what are the aspects which most spring to mind when scientific developments or discoveries are mentioned? MORI's 1999 survey on attitudes to bioscience for the British Government's Office of Science and Technology (using the 'People's Panel') began with a more general, unprompted, question (Table 1) to gauge precisely this.

It is therefore predictable that it is scientific developments related to advances in human health which spring most immediately to mind. At this unprompted question, as many as 41% of the public named at least one bioscience (increasing to 57% when treatments/cures for diseases was added in). Sixty-three % mentioned one or more non-bio science.

The main issues which the public would take into account in determining whether a biological development is right or wrong are whether people would benefit from it and whether it would be safe to use. Other significant considerations would be whether the benefits outweighed the risks, whether or not it interfered with nature, whether animals would be harmed and—something the question was in fact testing—whether it was considered to be right or wrong (Table 2).

Table 1. Q. Thinking about major scientific discoveries or developments, do any spring to mind?

	First mention %	Any mention %
Treatments/Cures for/Eradication of illnesses/diseases/ Medicines/New drugs/Penicillin/Antibiotics/Vaccines etc./Operations/Surgery	11	32
Computers/The Internet/Email/Millennium bug/ Compliance	8	21
Space/Sending people to the moon/Life on Mars	8	21
Genetically modified, genetically altered/genetically engineered food	9	18
Medical research	6	15
Cloning/Dolly the sheep	5	12
TV/New TV sets/Cable TV/Satellite TV/Digital TVs	4	12
New telecommunications (fax machine/mobile phone)	2	10
(Others below 10% omitted)		
No, none spring to mind	23	
Don't know	4	
Any mention	74	

Source: MORI/OST, 13–14 April 1999
Base: 1109 British adults

Thesis 2: Scientific developments aimed directly at achieving improvements in human health-care are the most valued by the public

The public have clear opinions that some scientific developments are beneficial and that others are not, as the OST survey shows. Again, advances in human health score highest, clearly representing in the public's mind the biggest benefit to arise from scientific developments. Specifically, the development of new medicines (antibiotics and vaccines) was most commonly mentioned, by 57% in the quantitative stage, followed by transplants of various organs (51%), cures for or eradication of diseases (43%), and new operations/surgery (31%). These medical benefits are clearly widely felt to be beneficial to society, and we found almost no advocates of a contrary view (Table 3).

Looking at 'net beneficial' scores (i.e. the proportion saying something is beneficial to society, minus the proportion saying it is not), we can see that development of new medicines receives a net score of +56 and transplants +50, cures for/eradication of illnesses +42, yet cloning gets a –55. Genetically modified food is also not regarded overall to be beneficial to society; it receives a net score of –44. Genetic modification

Table 2. Q. Now thinking about biological developments again, what things, if any, do you think you would personally take into account if you were deciding whether a particular development was right or wrong?

	First mention %	Any mention %
Whether it would help people/be beneficial	15	33
Whether it harmed people	10	29
Ethics/Morals/Whether it was right or wrong/Whether it was for the general good	12	22
Whether it had been tested properly/was safe	5	20
Whether the benefits outweighed the risks	8	19
Whether it had side-effects	3	17
Whether animals would be harmed	6	16
Whether it interfered with nature	5	16
Consideration for future generations	3	11
Whether it was controllable/well regulated	2	9
Whether it was useful to me/people	2	9
(Others 5% and below omitted)		
Don't know	16	
Any mention	82	

Source: MORI/OST, 13–14 April 1999
Base: 1109 British adults

of animals and plants also gets a negative score (of –27). In the qualitative phase of the research, group members made very little connection between animal cloning (viewed with suspicion) and advances in human health (the latter being of prime importance to the public).

One conclusion that seems clear from the OST survey is that scientific developments can gain quick acceptance if the public has wide experience of them and finds them useful. One area where public acceptance of scientific advances seems to be growing significantly is in information technology and telecommunications, especially the Internet, which scored a net +24 as beneficial in the survey for the OST.

A particularly clear instance of an ethical issue to be resolved in judging science is the question of experimentation on animals. The Jenkin Report says, 'Animal experimentation is a flashpoint in British society's relationship with science'. This was considered in considerably more detail in a survey for *New Scientist* (2000). Though this was an exploration of a specific scientific issue, it may be considered to have much wider implications as an exemplification of the way in which and extent to which the British public is prepared to trade off its ethical objections to scientific processes or developments in the light of the con-

Table 3. Q. On this card is a list of various scientific developments. Which two or three would you say have been *beneficial* for society as far as you are aware?
Q. And which two or three would you say have not been *beneficial* for society, as far as you are aware?

	Beneficial %	Not Beneficial %	Net ±%
Medicines/New drugs/Penicillin/Antibiotics/ Vaccines, etc	57	1	+56
Transplants, e.g. of heart, liver, kidneys, etc	51	1	+50
Cures for or eradication of illnesses/diseases	43	1	+42
New operations/Surgery	31	*	+31
Computers/The Internet/Email	28	4	+24
Genetic testing or screening for particular things, e.g. diseases	24	2	+22
Discovering global warming/Climate change/ Disruption to weather patterns/Greenhouse effect	19	6	+13
New and alternative sources of energy	17	4	+13
New telecommunications (fax machine/mobile phone/TV)	14	5	+9
Test-tube babies/*In-vitro* fertilization	11	9	+2
Faster/cheaper travel	6	16	−10
Robots in industry and medicine	3	18	−15
Splitting the atom	4	20	−16
Space research/Sending people to the moon	2	25	−23
Genetic modification/engineering of animals and plants	1	28	−27
Genetically modified food	1	45	−44
Cloning/Dolly the sheep	2	57	−55
Other	*	*	
No, none spring to mind	*	5	
Don't know	*	1	

Source: MORI/OST, 13–14 April 1999
Base: 1109 British adults

crete benefits to which they are intended to lead. We return to this later on in this paper.

Public attitudes towards animal experimentation are far from clear-cut. While, the initial 'knee-jerk' reaction may be one of opposition, the public is receptive to messages explaining (justifying) what benefits such experiments may bring. Two-thirds of the British public disagree that scientists should be allowed to conduct experiments on live animals; around a quarter agree, when asked in a 'cold-start' question, without any preface or prompting as to the purpose behind it or for the need to conduct such experimentation.

The purpose of an animal experiment has a significant effect on the public's likelihood, or not, to approve of it. The public differentiates substantially between curing leukaemia in children on the one hand and testing cosmetics on the other. Where no pain for the animal is involved, the balance of opinion is in favour of eight out of the nine experiments for mice, and seven out of the nine for monkeys.

As with the more general questions asked in the OST survey, the *New Scientist* research confirms that the public finds medically motivated research more acceptable than any other. The experiments that the public feels are most justified are those with a specific medical aim—for example, relating to curing leukaemia or AIDS. Experiments with less specific medical aims are not felt to be as justified, but are nevertheless felt to be more warranted than experiments to test garden insecticides and cosmetics. This 'ranking' remains identical regardless of the species to be experimented upon.

The public's perception of the purpose of scientific development also seems to affect their acceptance of procedures where their objection is probably principally a perceived health risk rather than ethical objections.

Thesis 3: Ignorance about the way in which science is regulated and restricted leads many of the public to assume that the regulation is insufficient, and this in turn makes them more likely to be hostile to science

Many of the public are unclear what scientists are trying to achieve in their work, which naturally reduces their understanding of the potential benefits and hence acceptance of the value of the research. This is especially true in the case of cloning which, as we have seen, is held by almost three in five of the public not to be beneficial to society (Table 3)

In the OST survey, three people in four, 77%, offered a reason why genetic modification takes place, or why there is GM food, 88% for why animal cloning takes place, and 66% for why 'transplants of animal tissues to humans' (xenotransplantation) takes place. However, some of the reasons perceived to be behind the research are unlikely to make it more publicly acceptable.

The public is aware that they do not know enough about scientific developments, and are keen to be better informed. In the workshops in the OST qualitative research, most expressed a desire for more information, many saying they had heard of Dolly 'after the event'. Discussions in the workshops about genetic modification often produced comment about the need for clear labelling of food.

A survey on public understanding of risk for the Better Regulation Office of the Cabinet Office explored in greater depth the relationship between perceived risk of scientific developments and self-assessed level of knowledge. MORI listed six possible health risks, and asked the public how well informed they felt about each, which they thought posed a serious threat to them or their family, and on which the government should legislate or provide advice and information about.

The general principle still stands: the better informed the public is by official and reliable sources, all other things being equal the more acceptable they are likely to find scientific development.

There is strong support for the government to be more open in its decision-making process. Over nine in ten think it should be more open (and 61% *strongly* agree), and eight in ten think that the government should release what information it does have even when it is unsure of the full facts (43% *strongly* agree). This (Fig. 1) reflects one of the most powerful findings from our qualitative study.

It is also clear that the public is keen to be given the facts to make up its own mind, even if the facts are not conclusive. An overwhelming 80% think that the government should publish what information it has available, even if it is unsure of the facts. This follows naturally from the public's distrust of the government generally, and to a lesser extent of government scientists, explored in detail below.

Q I am now going to read out a list of statements, and I would like you to tell me how strongly you agree or disagree with each.

The Govt. should do more to protect people by passing more laws that ban dangerous activites

When the Govt. is unsure of the facts, it should nonetheless publish what info. it does have available

The Govt. does not trust ordinary people to make their own decisions about dangerous activities

I am confident the Govt. will act generally in the public interest

The Govt. should be more open about how it makes its decisions

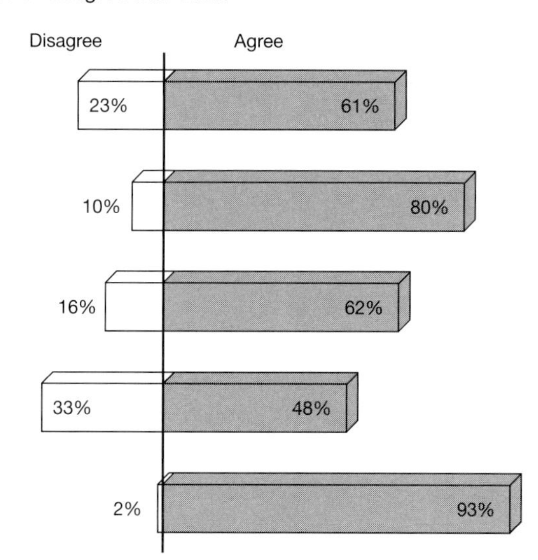

Fig. 1.

Thesis 4: The majority of the public say they trust 'scientists' but whenever a scientist's employer or sponsor is mentioned, the veracity of the source becomes highly relevant: the scientists trusted by the highest proportion of people are those working for environmental NGOs

It is not that the British public is unusually sceptical. In a six-country study (Table 4) several years ago, we were asked by the Pew Center in the USA to carry out fieldwork on a study of trust here in Great Britain, while others replicated the work in their countries, Germany, Spain, and France, and Pew in the USA. Somewhat to our surprise, we found that significantly more of the British public were willing to trust the state here, than elsewhere in the countries under examination. While a majority, 57%, of the British said that they basically trust the state in this country, only four in ten Americans and Germans were willing to do so, 38% of the Spanish, and only about a third of the Italians and the French.

If the public distrusts what scientists say, it is a barrier both to their willingness to listen to scientific arguments for the need for experimentation (or to the case for considering alternatives) and to arguments that humankind will suffer and/or die for the lack of it. Such considerations are particularly pertinent in cases such as the question of animal experimentation, where there is widespread and highly emotionally charged 'instinctive' opposition to the idea, and a reasoned case needs not only to be made out but to be put across to the public if the work involved is to find any degree of general acceptance. In such cases it is not realistic to assume that the public will hold an open mind.

There is considerable scepticism among the public about scientists' competence as experts. In the field of the environment, this perception has been fairly steady for a number of years, with the public generally fairly evenly divided on the proposition that 'Even the scientists don't really know what they're talking about when it comes to the environment' (Table 5). This perception, of course, is fostered by the way in

Table 4. Q. Would you say you basically trust the state or not?

	Yes %	No %	DK %
Great Britain	57	32	11
Germany	41	41	18
USA	40	56	4
Spain	38	42	20
Italy	35	51	14
France	33	59	8

Source: PEW
Base: 1762 in USA, *c.* 950 in other countries (1997)

Table 5. Q. Do you agree or disagree? 'Even the scientists don't really know what they're talking about when it comes to the environment'

	1989 %	1992 %	1993 %	1995 %	1996 %	1997 %	1998 %	1999 %
General public								
Agree	36	37	41	36	n/a	49	41	48
Disagree	37	38	34	35		33	37	32
Net agree	−1	−1	+7	+1		+16	+4	+16
Environment journalists								
Agree	n/a	n/a	19	21	31	16	n/a	n/a
Disagree			65	63	62	60		
Net agree			−46	−42	−31	−44		

Source: MORI Annual Business and the Environment study and Environment Journalists survey (Don't knows omitted)
Base: *c.* 1000 British adults each year, *c.* 25 environment journalists each year

which the media, especially the broadcasting media, cover scientific news stories. Conflicting scientific reports abound, and public understanding is not helped by the media's habit of 'even-handedness', pitting the spokesperson for say 'global warming', who represents the vast majority of scientific opinion, against the scientific sceptic, who speaks for a few dissidents. The Jenkin Report authors quibble about the wording of the question, derived from a statement made in a focus group, saying that it may be interpreted by different people in different ways, for instance that some people might take it to mean that 'Environmental researchers are incompetent', rejecting this outright, saying 'This is patently not true'. Oh? If we were to employ a Lickert agree–disagree question of the general public, I would bet that there would be those who would assert that this is indeed their *perception*, and it is perceptions that their report was supposed to be about.

On the whole, this does not give an accurate picture of what the environmental journalists themselves think—when asked the same question about scientists' knowledge, the majority usually disagree. Nevertheless, in MORI's 1996 Environment Journalists survey, the number agreeing that scientists didn't know what they were talking about was as high as three in ten. More to the point, the Report states that scientists must learn to work with the media as they are.

The perception is of course swayed by events, and in 1997 the margin in favour was considerably greater than has been the trend. Such public failures as the government scientists' U-turn over BSE and the embarrassment of Greenpeace's scientists over Brent Spar presumably contributed. In the circumstances it would not be surprising if the public were wary of taking any single scientific pronouncement as representing authoritative truth.

Nevertheless, the majority of the public generally trust scientists to tell the truth. In general, scientists perform reasonably well (but not outstandingly) when compared with other groups on how far the public trusts them to tell the truth. This can be tested in two ways. MORI's standard 'veracity' test (last conducted for the British Medical Association in February 2000) asks respondents to judge for each group whether they are generally trusted or not; in both 1997 and 1999 63% and in 2000 60% of the public said they trusted scientists, putting them ahead of the benchmark figure (56% in 1997 and 52% in 2000) of 'the ordinary man/woman in the street', but well behind the most trusted groups, doctors and teachers, and indeed behind professors (Table 6).

By way of contrast, when the Louis Harris polling organization asked an identical question in the USA in 1998, scientists came near the top of the list, trusted to tell the truth by 79% of the American public, this despite the recent finding by Dr Bodenheimer of the University of California at San Francisco who reported in an article in the *New England Journal of Medicine* that when drug companies paid for a trial of a new drug, 89% of the time the studies found that the new drug was

Table 6. Q. Now I will read out a list of different types of people. For each, would you tell me whether you generally trust them to tell the truth or not?

	Tell the truth %					Not tell the truth %				
	'83	'93	'97	'99	'00	'83	'93	'97	'99	'00
Doctors	82	84	86	91	87	14	11	10	7	9
Teachers	79	84	83	89	85	14	9	11	7	10
Television news readers	63	72	74	74	73	25	18	14	17	18
Professors	n/a	70	70	79	76	n/a	12	12	10	11
Judges	77	68	72	77	77	18	21	19	16	15
Clergyman/Priests	85	80	71	80	78	11	13	20	14	16
Scientists	*n/a*	*n/a*	*63*	*63*	*60*	*n/a*	*n/a*	*22*	*27*	*25*
The Police	61	63	61	61	60	32	26	30	31	33
The ordinary man/ woman in the street	57	64	56	60	52	27	21	28	28	34
Pollsters	n/a	52	55	49	46	n/a	28	28	35	35
Civil servants	25	37	36	47	47	63	50	50	41	40
Trade union officials	18	32	27	39	38	71	54	56	47	47
Business leaders	25	32	29	28	28	65	57	60	60	60
Journalists	19	10	15	15	15	73	84	76	79	78
Politicians generally	18	14	15	23	20	75	79	78	72	74
Government ministers	16	11	12	23	21	74	81	80	70	72

Source: MORI/British Medical Association 1999, 2000; *Times* 1993, 1997; *Sunday Times* 1983
Base: *c.* 2000 British adults

better than ones it was replacing, but when tests were done by scientists who were not paid by the drug companies, the new drug received good marks only 61% of the time. Yet, the scientists' 'veracity rating' in the USA is 16 points higher than in the UK. In general Americans tend to rate their professions much higher than the British, with the conspicuous exception of their TV newsreaders. While three in four of the British say they generally trust their newsreaders to tell the truth, fewer than half of the Americans do; also, it is interesting to see that a much higher percentage of Americans rate their journalists as trustworthy, but the most startling finding in my view, and worth pondering, is that nearly double the number of Americans say they trust their civil servants (70%) than do the British (36%)—Table 7.

Just as the public trusts some groups in society more than others to 'tell the truth' in general, so, also there is a hierarchy of trust in providing 'honest and balanced information' about specifically scientific matters, which combines requirements of having the accurate knowledge and integrity in reporting it.

Again, doctors top the list, but on this question there is a much clearer division into several categories—what might be described as the profes-

Table 7. Q. Now I will read out a list of different types of people. For each, would you tell me whether you generally trust them to tell the truth or not?

	1997 (GB)			GB/USA	1998 (USA)		
	Trust %	Not %	Net ±%	±Trust %	Trust %	Not %	Net ±%
Doctors	86	10	+76	−3	83	13	+70
Teachers	83	11	+72	+3	86	13	+73
Clergymen/priests	71	20	+51	+14	85	13	+72
Professors	70	12	+58	+7	77	19	+58
Judges	72	19	+53	+7	79	18	+61
Television newsreaders	74	14	+60	−30!	44	52	−8
Scientists	63	22	+41	+16	79	18	+61
The police	61	30	+31	+14	75	23	+42
The ordinary man/woman in the street	56	28	+28	+15	71	25	+46
Pollsters	55	28	+27	0	55	38	+17
Civil servants	36	50	−14	+34!	70	27	+48
Trade union officials	27	56	−29	+10	37	58	−21
Business leaders	29	60	−31	+20	49	47	+2
Government ministers	12	80	−68	n/a	46	51	−5
Politicians/The President	15	78	−63	n/a	54	44	+10
Journalists	15	76	−61	+28!	43	52	−9

Source: MORI/Times (1997)/Louis Harris & Associates
Base: *c.* 1000 British adults/*c.* 1013 American adults

sional experts, including 'scientists', are trusted by more than distrust them. 'The general public' and 'patients' both score close to zero, though with marginally negative net scores. Other groups, however, ranging from farmers to animal welfare groups to industry to religious groups are greatly more distrusted than trusted. But both consumer groups and environmental groups score well (Table 8).

The media is greatly distrusted, with more than three times as many respondents saying they would *not* trust the media to provide them with honest and balanced information as say they would trust them. The media—especially television—is the public's main source of informa-

Table 8. Q. Which, if any, of the following types of people or institutions would you trust to provide you with honest and balanced information about biological developments and their regulation?
Q. And which, if any, would you *not* trust to provide you with honest and balanced information about biological developments and their regulation?

	Trust %	Not trust %	Not trust ±%
GPs/Family doctors	60	6	+54
An advisory body to government, composed of people representing different viewpoints	48	9	+39
Hospital doctors	44	6	+38
An advisory body to government, composed of experts	47	15	+32
Pharmacists/Chemists	32	10	+22
Consumer groups	33	15	+18
Nurses	23	6	+17
Scientists	34	20	+14
Vets	20	8	+12
Environmental groups	31	20	+11
The general public	14	17	−3
Patients	6	11	−5
Sociologists	9	16	−7
Animal welfare groups	16	35	−19
Governments	19	39	−20
Religious organizations	9	32	−23
Farmers	8	34	−26
The media	14	47	−33
Retailers	4	46	−42
Industry/Manufacturers	3	53	−50
None of these	2	1	
Don't know	1	2	

Source: MORI/OST, 13–14 April 1999
Base: 1109 British adults

tion. Inevitably, this must mean that the public is inclined to treat much that it hears about science with suspicion.

These perceptions have changed little over the years. A 1985 MORI survey for the Technical Change Centre asking which groups from a list of sixteen respondents would, and which they would not, 'trust to tell the truth about the effect scientific and technological developments will have on our lives', found very similar results. Doctors topped the poll while Members of Parliament and newspaper journalists were most distrusted (by 63% and 66%, respectively), and consumer groups scored better than environmental groups. Interestingly, 'presenters of scientific programmes on television', while trusted by 37% were also distrusted by a substantial 23%. Perhaps more significantly, however, twice as many distrusted 'scientists working for major companies' as trusted them.

Some scientists are trusted more than others. MORI surveys have persistently found that trust in scientists' pronouncements are affected by knowledge of who is sponsoring the scientists' research. A 1997 survey for the Cancer Research Campaign (Table 9) asked the public how much confidence they would have in what each of ten groups had to say about their research projects and findings. The ten groups listed were four cate-

Table 9. Q. How much confidence would you have in what each of the following have to say about their research projects and findings?

	A great deal %	A fair amount %	Not very much %	None at all %	Don't know %	A great deal/ fair amount %
Cancer research charities	39	49	6	1	4	88
Medical research charities	37	50	6	2	5	87
Mental health charities	31	53	8	2	6	84
Animal welfare charities	20	53	18	5	5	73
Scientists working for the IT (computer) industry	15	53	16	4	12	68
Scientists working for industry	6	53	30	5	6	59
Scientists working for the government	5	45	38	6	5	50
Scientists working for the chemical industry	7	35	41	10	7	42
Scientists working for the nuclear industry	6	26	41	20	7	32
Scientists working for the tobacco industry	3	15	42	34	6	18

Source: MORI/Cancer Research Campaign 9–12 May 1997
Base: 1933 British adults

gories of charity and scientists working for six different types of sponsor. All four charitable categories scored higher trust ratings than any of the scientists.

But, more significantly, there were very substantial differences in reactions to the scientists. At one end of the scale, two-thirds of the public (68%) had at least a 'fair amount' of confidence in what 'scientists working for the IT (computer) industry' said; at the other, barely one in six (18%) trusted scientists working for the tobacco industry to the same degree.

The CRC research is not an isolated finding. MORI's Business and the Environment studies regularly test trust in different groups of scientists on the more specific question of what they have to say about environmental issues. The surveys invariably find that the public have considerably more confidence in what 'scientists working for environmental groups' have to say about environmental issues than 'scientists working in industry', who in turn have tended to be slightly more trusted than 'scientists working for the government', although they are about the same in the last (1999) survey—Table 10.

MORI's surveys of environment journalists, asking the same question, have also found a clear hierarchy of trust; but in their case, scientists working in industry are very much less trusted than those working for the government. In 1998, 83% of the environment journalists said they had a great deal or a fair amount of trust in scientists working for environmental groups, 67% in those working for the government, and just 29% in those working in industry.

A separate study, the survey for the Cabinet Office's Better Regulation Unit, found a similar differentiation of trust. The survey went on to examine two specific environmental issues, pollution and BSE (see Figs 2 and 3); while 'independent scientists' scored a high degree of trust, 'government scientists' are very much more distrusted.

On the subject of pollution, the public has most confidence in pressure groups and 'independent scientists', each trusted to give advice by three-fifths. Both are especially likely to be picked out by those with profes-

Table 10. Q. How much confidence would you have in what each of the following have to say about environmental issues?

	A great deal/a fair amount				
	1995	1996	1997	1998	1999
	%	%	%	%	%
Scientists working for environmental groups	82	75	83	75	78
Scientists working in industry	48	45	47	43	48
Scientists working for the government	38	32	44	46	47
Net lead (environmental groups over government)	–42	–43	–39	–39	–41

Source: MORI Annual Business and the Environment studies
Base: *c.* 2000 British adults each year

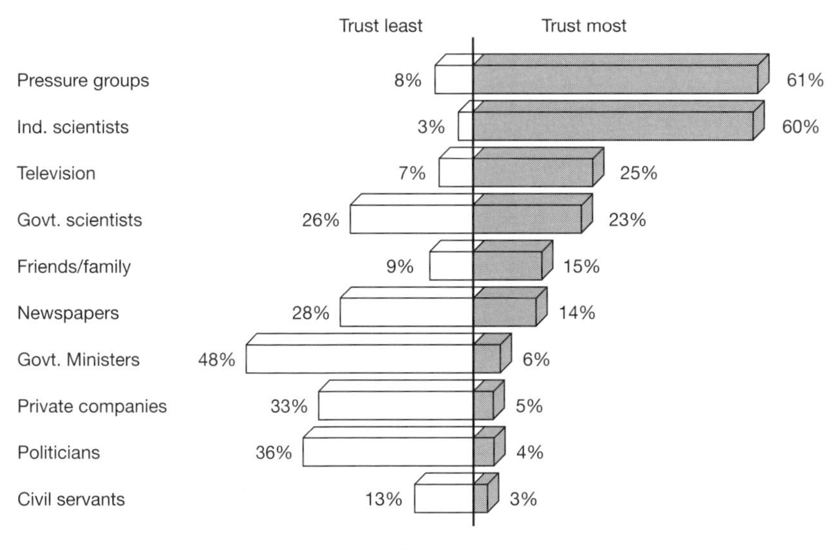
Q Thinking now about pollution, which two or three, if any, of these sources would you trust most/least to advise you on the risks of pollution?
Trust least
Trust most
Pressure groups
8%
61%
Ind. scientists
3%
60%
Television
7%
25%
Govt. scientists
26%
23%
Friends/family
9%
15%
Newspapers
28%
14%
Govt. Ministers
48%
6%
Private companies
33%
5%
Politicians
36%
4%
Civil servants
13%
3%

Fig. 2.

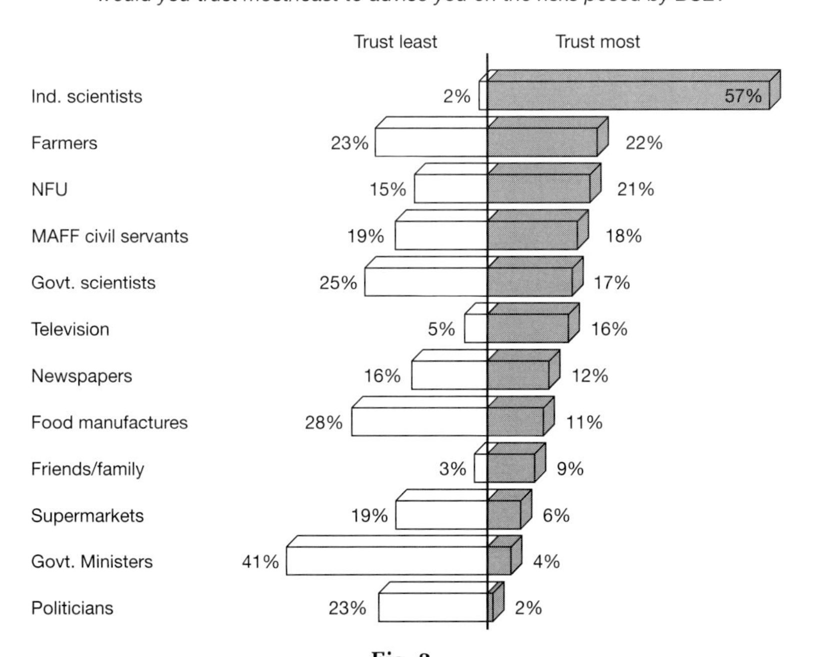
Q Thinking now about BSE, which two or three, if any, of these sources would you trust most/least to advise you on the risks posed by BSE?
Trust least
Trust most
Ind. scientists
2%
57%
Farmers
23%
22%
NFU
15%
21%
MAFF civil servants
19%
18%
Govt. scientists
25%
17%
Television
5%
16%
Newspapers
16%
12%
Food manufactures
28%
11%
Friends/family
3%
9%
Supermarkets
19%
6%
Govt. Ministers
41%
4%
Politicians
23%
2%

Fig. 3.

sional jobs, and less likely by those aged 65+. Interestingly, television is trusted much more than newspapers—reflecting one of the findings of our qualitative work.

Although this differentiation of trust levels depending on who employs a scientist may be felt to be fairly case-specific, its implications are much wider. It is fairly obvious, of course, in the context of the surveys quoted, that the public's trust of each group of scientists is directly related to its trust of the scientists' employers. However, this clearly carries the implication that, consciously or unconsciously, the public rejects either the idea that scientists are capable of objective and reliable research or the idea that they can be trusted to tell the truth about it if it is not in their employers' interests to do so. This puts more general measures of trust in 'scientists' into disturbing context: a generally reasonable level of public trust in scientists may not reflect any belief in their integrity, but merely a lack of any perception that 'scientists' in general (as opposed to specific groups of employed scientists) would be under any pressure to falsify results. If this is so, underlying mistrust of scientists may be higher than it seems at first.

British public attitudes to animal experimentation

When approached by the *New Scientist* magazine to conduct a study into British attitudes on experimentation of live animals, it became clear that there were multiple objectives to be tested on British public opinion. These included:

1. Public support for or opposition to animal experimentation generally, testing by use of a 'split-sample'/'split-ballot' test for differences between public support/opposition to animal experimentation to a 'cold-start' question, asking simply whether or not scientists should be allowed to conduct any experiment on live animals, or a 'warm-start' question, prefaced by a description of the end use of the experimentation, such as reducing pain or developing new treatments, the hypothesis being that public support would be greater, i.e. more people would support animal experimentation if reminded what it was being done for.
2. Public differentiation between various end applications for animal testing, including comparisons of support for or opposition to animal testing depending on whether the end objectives of the scientific experimentation was to test for safety and efficacy, or to develop cures for disease, the hypothesis being that more people would support animal experimentation in aid of basic medical research to find new cures, rather than for safety and efficacy. A variety of actual live animal experimentation was tested, with the addition of 'anchor

points', cosmetic testing (no longer allowed in Britain) at the one extreme, and finding a cure for leukaemia in children at the other.
3. Public belief as to the comparative animal testing regulatory stance taken in Britain v. other developed countries, the hypothesis being that the British people, while not knowing or understanding the details of animal experiment regulations in Great Britain, would be more likely think that the British, being thought an animal-loving people, would have stricter regulations.
4. Public differentiation between the type of animals used in experimentation, e.g. mice v. monkeys, the hypothesis here being that mice bred for experimentation were nonetheless not seen as so close to human, were rodents, and were short-lived and not so 'cuddly', and would therefore be preferred as experimental animals.
5. Public discrimination between types of end objective, causing pain v. death of the animal used in the experimentation, the hypothesis being that pain for the animals would be seen by more of the British public as preferable to causing termination.
6. Finally, using the facility of the survey to determine the incidence and profile of the British public's 'scientific behaviour', from taking part in demonstrations against animal experimentation to visiting the countryside, with the parallel opportunity to cross-tabulate the behavioural data with the attitudinal data on animal experimentation.

Working from the Opinions/Attitudes/Values model, we believed that we would find that for some people, animal experimentation will be a deep, strongly held, value, learned by their childhood experiences and for these people no argument would persuade them of the need for what they see as unreasonable cruelty to animals, causing them pain, suffering, and death. Others, perhaps those whose lives or the lives of their families have been saved by drugs, such as penicillin, where a key part of the experimentation leading to the development and/or adoption of the drug involved the death of animals, would feel as strongly.

Another factor that we felt had to be taken into account when determining public acceptability of animal experimentation was their trust in scientists. If the public distrusts what scientists say, either because of their own scepticism, or experience, or controversy, or confusion, it is a barrier both to willingness to listen to scientific arguments for the need for scientific experimentation on live animals, or to the case for considering alternatives, and the likelihood that humankind will suffer and/or die for the lack of it. It is important to remember that while generally trust in scientists to tell the truth is relatively high (63% said at the time the study was being designed that they felt they can trust scientists to tell the truth, and doctors, at 91% top the poll), when asked specifically who they trust to tell the truth about the environment, 75% say they trusted what scientists working for environmental groups have to say about environmental issues, just 46% say they trusted scientists working for the government, and 43% say they trusted what scientists working in industry say.

We found

- British public opinion is largely against experimentation on live animals. Two-thirds of the British public disagree that scientists should be allowed to conduct *any* experiments on live animals (the word 'any' expressly inserted to make the statement as 'tough' as we possibly could, measuring values, rather than attitudes or opinions), around a quarter of the British public, which would project to over 10 million British adults, agreed 'strongly', when asked in a 'cold-start' question and without any preface or prompting as to the purpose behind it or for the need to conduct such experimentation, in a 'split-ballot' experiment. On the other side, only 4% said they felt 'strongly' that scientists should be allowed to carry out any experiments on live animals.
- Public attitudes towards animal testing (unprompted) vary across different groups in society—those groups more opposed to animal experimentation include:
  - women, compared with men (71% of women disagreed with animal testing versus 57% of men);
  - 15–54-year-olds, four in five of the adult population, compared with those aged 55+ (67% versus 58%);
  - Labour party supporters, compared with Tories (69% versus 55%).
- Unlike much of MORI's public attitudes work, though, there are slight differences in opinion across the social grades, although while 69% of ABs, the professional and managerial classes, opposed live animal testing, ten points fewer, 59%, of skilled workers opposed it (59%).
- 'Lifestyle factors', the behavioural activities of people, reveal a number of attitudinal differences—suggesting a degree of correlation between people's 'respect' (used loosely) for the environment generally and attitudes towards animal experimentation, although the low base sizes mean that some of the following findings should be treated with caution:
  - disagreement with animal experimentation was highest among vegetarians (85%, base: 80; 9% of adults), members of animal welfare organizations (83%, base 79), those who have signed a petition on animal welfare (86%, base 204), and those who have bought cruelty-free products (77%, base 308);
  - conversely, opposition is comparatively lower among those who fish (62%), have studied a scientific subject (61%), and who say they have refused to take a drug for a serious illness that had been

tested on animals (52%, base: 65 people), but note that all these are on balance against animal testing;
- most strikingly, though, 38% (some 26 points less than the public as a whole) of those who have taken part in blood sports and/or worn a fur coat (base: just 23 people) disagree with animal experimentation—the majority, 62%, agree

When the motivations and possible outcomes for such experiments are explained to the public (i.e. the question is deliberately 'biased'—the 'biased' and 'non-biased' questions were asked of *separate* but matched samples of *c.* 1000 adults aged 15+), their opposition towards animal experiments crumbles by 23 percentage points to four in ten (41%). Thus, when the 'stakes are upped' in this way, the balance of opinion (albeit marginally) shifts *in favour* of live animal tests, with 45% agreeing that scientists should be allowed to conduct them, resulting from a 'swing' (in political terms) of 22%, twenty-two people in a hundred having been persuaded to change their minds on the issue by the preface to the question, explaining that 'Some scientists are developing and testing new drugs to reduce pain, or developing new treatments for life-threatening diseases such as leukaemia and AIDS. By conducting experiments on live animals, scientists believe they can make more rapid progress than would otherwise have been possible', before asking the agree/disagree question put to the other half of the split sample.

- Where the public is exposed to a description of the alleged benefits of such experiments, some of the subgroup differences observed at the 'non-biased' question still persist. Those groups more likely to disagree with animal experimentation in this instance are:
  - women, compared with men (46% versus 36%);
  - Labour, compared with Conservative supporters (42% versus 29%).
- There are significant differences by age group, with the youngest, 15–24-year-old segment still hostile to animal testing (–14) while the middle-aged, 35–54, are on balance positive (+12); differences here also do exist across the social grades:
  - C_2DEs are on balance against (–6), compared with ABC_1s who are on balance favourable to animal testing, when prompted (+14).
- As the changes in subgroups compared with the 'non-biased' question suggest, the act of 'loading' the question has more effect on some subgroups than others—serving to create significant differences in some cases, and to cancel them out in others. Agreement among ABs and C_1s falls by 34 and 26 percentage points respectively, while agreement among C_2s and DEs falls much less—by 16 and 17 points, respectively. This resulted in the significant difference noted above.
- On the other hand, disagreement among 35–44-year-olds and 45–54-year-olds falls by 29 points and 30 points, respectively—more so than

among the other age groups, particularly those aged 65+. This serves, to somewhat cancel out the age-group differences observed at the 'non-biased' question.

- For the most part, the lifestyle group differences remain when the question is 'loaded'. Disagreement with animal experimentation is highest among vegetarians (56%, base: 80), members of animal welfare organizations (68%, base: 70), and those who have signed a petition on animal welfare (55%). While they stood out at the 'non-biased' question, cruelty-free product buyers in this instance are more in step with the population as a whole.
  - conversely, again, disagreement is comparatively lower than among those who have studied a scientific subject (32%) and those who say they have refused to take a drug for a serious illness that had been tested on animals (25%, base of 60 people);
  - most particularly, though, again, just 12% of those who have taken part in blood sports and/or worn a fur coat (base: 26) disagree with animal experimentation—the majority (71%) agree

As far as they know, two-thirds of Britons consider that our regulations for controlling animal experiments are stricter or about the same as in most other developed countries—including four in ten saying they think they are stricter. Just one in ten think British regulations are less strict.

- Those groups more likely to consider British regulations to be stricter than elsewhere are:
  - men, compared with women (45% versus 37%);
  - those aged 35+, compared with 15–34-year-olds (43% versus 36%);
  - ABC_1s and C_2s combined, compared with DEs (44% versus 34%).
- There is a degree of overlap between those groups who are less opposed to animal testing and those who think that Britain's regulations are stricter than elsewhere. Nevertheless, as with the much quoted 'chicken and egg' example, we cannot infer from the data whether attitudes towards animal experimentation reflect perceptions of how such tests are regulated or vice versa.
- It must also be borne in mind that, to some extent, the subgroup variations above, reflect greater or lesser proportions of these groups answering 'don't know', rather than actively considering the regulations to be about the same or not as strict elsewhere. Women, for example, are seven points more likely to say 'don't know' at this question than men—which is very similar to the percentage difference between men and women considering British regulations to be stricter. Arguably, building on the discussion on the effect of loading the general attitude question, it is possible that increasing public awareness of the stringency of UK regulations might help create a more favourable attitude to such experiments.

- Lifestyle factors have notably less of an effect on perceptions of stringency of regulations than they do on attitudes towards the experiments themselves—the difference between the highest and lowest subgroup findings is 15 points, compared with 39 points for the 'non-biased' question.
- The lifestyle groups most likely to consider Britain's regulations to be stricter than elsewhere include those who have refused a drug tested on animals (57%) and fishing (48%) (in line with their being less likely to disagree with animal testing generally).
- Despite not standing out particularly at the attitudes towards experimentation questions, those who have been to a farm (70%), studies a scientific subject (70%), or read a science magazine (69%) are also more likely than other groups to consider Britain's regulations to be stricter than elsewhere. This does, perhaps, cast some doubt on how directly interlinked these two elements are (experimentation and regulation).
- Once again, the strongest views are held by those who have taken part in a blood sport or worn a fur coat (57%, base: 49).

The purpose of an animal experiment has a significant effect on the public's likelihood, or not, to approve of it. The public differentiate substantially between curing leukaemia in children on the one hand and testing cosmetics on the other—Figs 4 and 5 show, where no pain is involved, how net support falls from statement H to A—a fall of exactly 91 percentage points for both mice and monkeys.

- Where no pain is involved, the balance of opinion is in favour of 8 out of the 9 experiments for mice, and 7 out of the 9 for monkeys.
- The experiments that the public feels are most justified are those with a specific medical aim—for example, relating to curing leukaemia or AIDS. Experiments with less specific medical aims are not felt to be as justified, but are nevertheless felt to be more warranted than experiments to test garden insecticides and cosmetics.
- As the charts show, this 'ranking' remains identical regardless of the species tested.
- The second most disapproved-of experiment, for both species, is to test whether a garden insecticide will be harmful to people. This is of particular concern given that this type of experiment accounts for a good proportion of animal experiments that are actually carried out.

The public does not differentiate significantly between efficacy testing of drugs and vaccines and their development—across either species.

Also, the public does not differentiate greatly between causing pain or death in either species—indeed, in some instances, the observed approval figures are marginally higher for where pain is suffered than

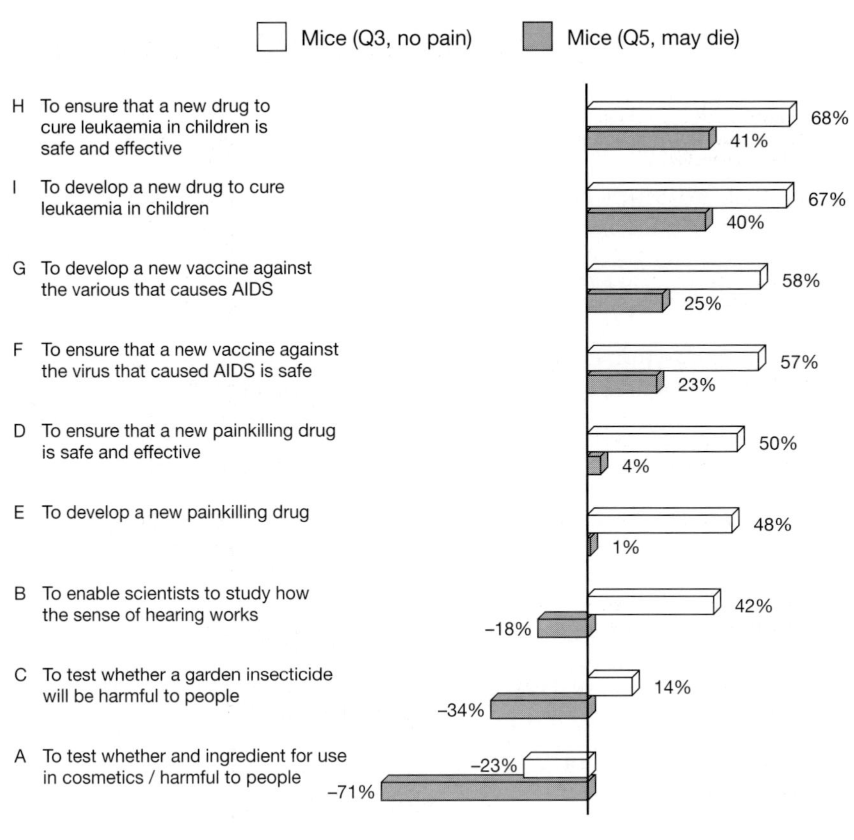

Fig. 4.

possible death. For this reason, to simplify the charts included in this report, they incorporate the 'no pain' and 'possible death' data, and exclude the 'suffering pain' data. We would suggest, that the public does not distinguish between the two, as arguably the words 'in pain' are more emotive and imply more suffering than 'death' itself—which might be seen as a 'release'.

- As well as differentiating between experiment purpose, Figs 4 and 5 show that the approval of each experiment drops noticeably between when no pain is involved and when the animals may die. For experiments H, I, G, and F, for mice, the differential between 'no pain' and 'possible death' is around 27–34 percentage points. Where the outcome is less specific—experiments D and E—the differential increases to 46/47 points, a similar differential as for experiments C and A (though notably fewer approve of C and A in the first instance). The most striking fall, though, is for experiment B—for which net approval drops by 60 points between 'no pain' and 'possible death'.

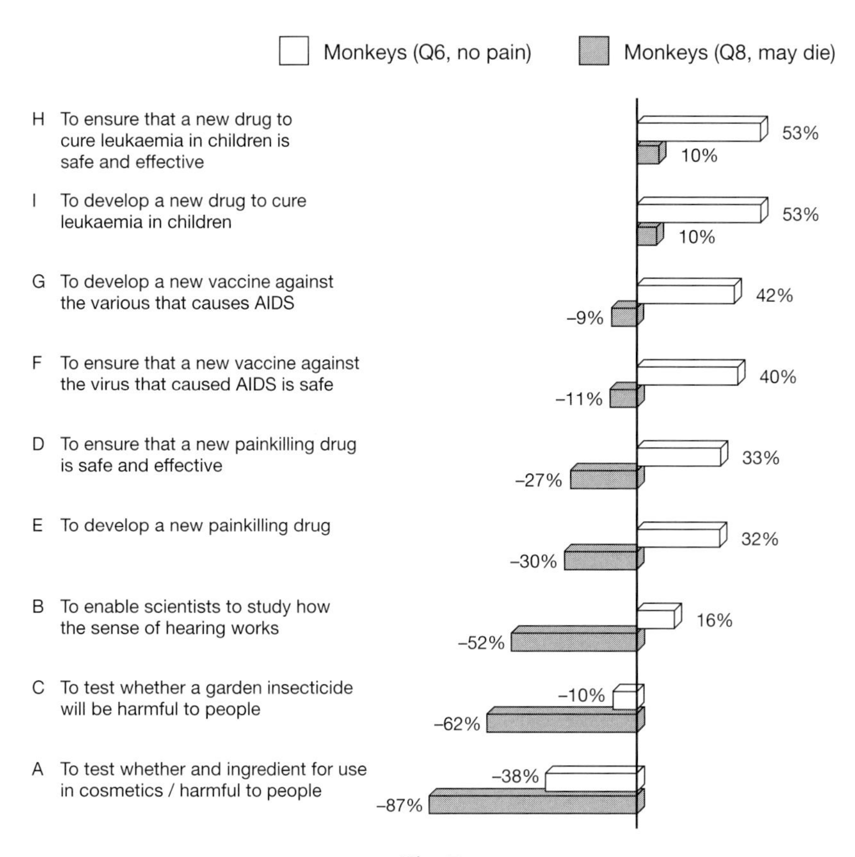

Fig. 5.

- The pattern by severity is similar for monkeys—though not identical. First, the differentials between no pain and pain are notably higher than the differentials for mice. The smallest differentials for monkeys, as for mice, are for statements I, H, G, and F—though, for monkeys, the public does distinguish between the two sets I and H (differential of 43 points) and G and F (differential of 51 points).
- As with mice, the greatest differential is for experiment C—testing whether a garden insecticide would be harmful to people—with 72 points difference. The next widest differential is for studying the sense of hearing (68 points). The differential is not as wide for cosmetic testing—but again, support is notably lower for this experiment in the first place.
- Figures 6 and 7 show that the public does differentiate between using mice and monkeys for each of the experiments—and substantially so. Where the animals do not suffer pain, illness or surgery (Fig. 6), there is a difference in net approval of around 15–17 points between mice and monkeys. For two experiments though, B and C, this differential is

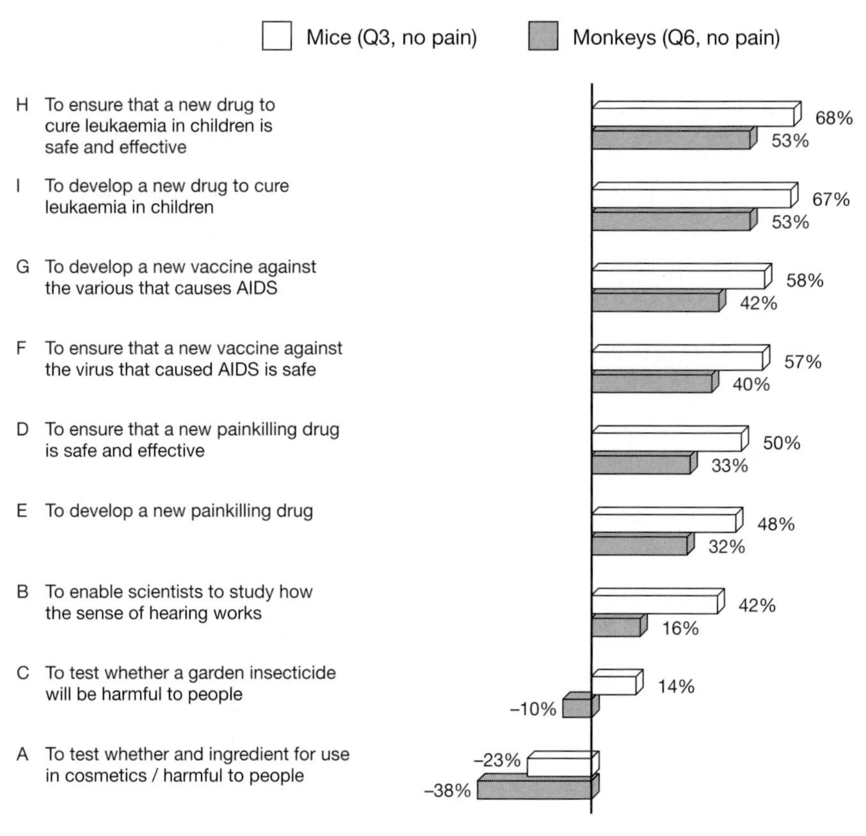

Fig. 6.

wider—around 25 points. The inference is that the public is even less prepared to accept these experiments where monkeys may die than where mice may do so. Where the animals may die (Fig. 7), the differential between mice and monkeys is a fairly consistent 28–34 points—reflecting the higher value the public places on the lives of monkeys, compared with those of mice.

- The differential between mice and monkeys for cosmetic testing is notably lower than for the other experiments—however, this reflects a high level of disapproval for the use of both animals, rather than that a greater relative approval of the use of monkeys.
- If the net support for each of the experiments on both mice and monkeys, where the animals do not experience pain, are plotted (mice on the x axis, and monkeys on the y), the points are highly correlated ($r^2 = 0.98$). This illustrates the earlier point that the experiment ranking remains the same for each species.
- If the same graph is plotted for where the animals experience pain and may die, the high correlation remains. However, the line of best fit

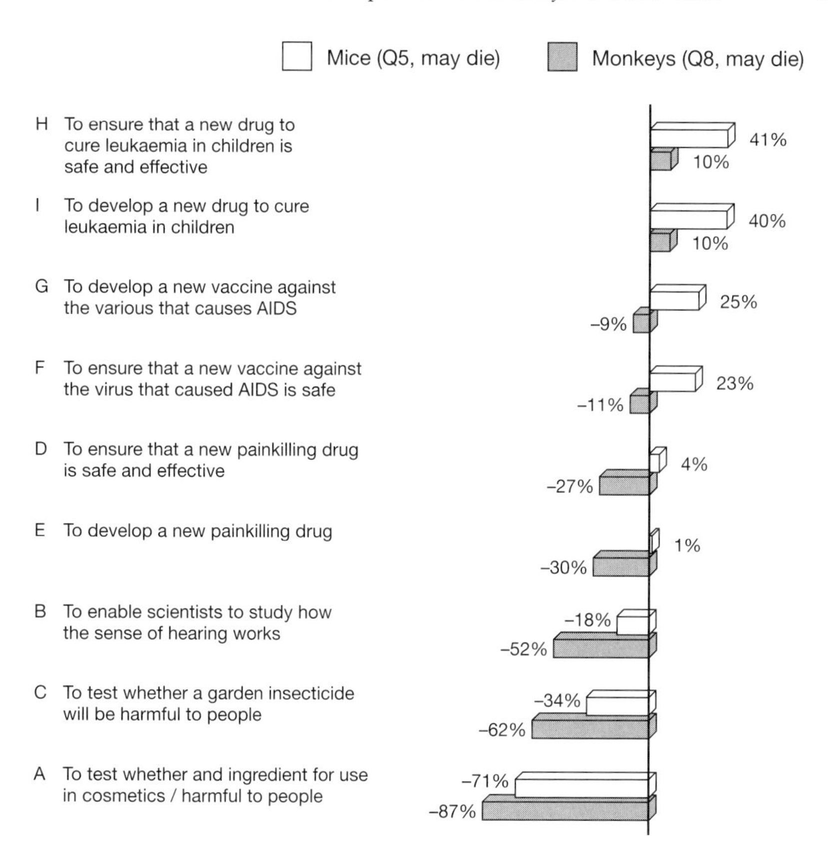

Fig. 7.

shifts along the *x* axis (from 18 to 32)—illustrating the earlier point that as the degree of pain increases, the public is even less prepared to accept experiments on monkeys than they are on mice.

Table 11 summarizes the findings from the various split-ballot experiments, comparing the use of live mice and live monkeys in experiments conducted by scientists for a variety of end purposes, to determine the degree of public support/opposition depending on which of the two species is being employed in the experiments, and the end use, across a spectrum ranging from use in testing cosmetics to seeking a cure for leukaemia in children.

The alternatives tested were:

A. To test whether an ingredient for use in cosmetics will be harmful to people.
B. To enable scientists to study how the sense of hearing works
C. To test whether a garden insecticide will be harmful to people
D. To ensure that a new pain-killing drug is safe and effective

Table 11. Comparative analysis of use of mice and monkeys in live animal experiments

Mice v. Monkeys	A	B	C	D	E	F	G	H	I	MEAN	STDDEV
Monkeys (no pain)	–0.38	0.16	–0.10	0.33	0.32	0.40	0.42	0.53	0.53	24.6%	0.306
Mice (no pain)	–0.23	0.42	0.14	0.50	0.48	0.57	0.58	0.68	0.67	42.3%	0.293
Difference	–0.15	–0.26	–0.24	–0.17	–0.16	–0.17	–0.16	–0.15	–0.14	–17.8%	0.042
Monkeys (pain, illness or surgery)	–0.85	–0.54	–0.65	–0.27	–0.25	–0.08	–0.08	0.08	0.08	–28.4%	0.330
Mice (pain, illness or surgery)	–0.74	–0.25	–0.39	–0.02	–0.03	0.17	0.18	0.33	0.33	–4.7%	0.358
Difference	–0.11	–0.29	–0.26	–0.25	–0.22	–0.25	–0.26	–0.25	–0.25	–23.8%	0.051
Monkeys (may die)	–0.87	–0.52	–0.62	–0.27	–0.30	–0.11	–0.09	0.10	0.10	–28.7%	0.331
Mice (may die)	–0.71	–0.18	–0.34	0.04	0.01	0.23	0.25	0.41	0.40	1.2%	0.370
Difference	–0.16	–0.34	–0.28	–0.31	–0.31	–0.34	–0.34	–0.31	–0.30	–29.9%	0.056
Monkeys											
May Die	–0.87	–0.52	–0.62	-0.27	–0.30	–0.11	–0.09	0.10	0.10	–28.7%	0.331
No Pain	–0.38	0.16	–0.10	0.33	0.32	0.40	0.42	0.53	0.53	24.6%	0.306
Difference	–0.49	–0.68	–0.52	–0.60	–0.62	–0.51	–0.51	–0.43	–0.43	–53.2%	0.085
Pain, Illness or surgery	–0.85	–0.54	–0.65	–0.27	–0.25	–0.08	–0.08	0.08	0.08	–28.4%	0.330
No Pain	–0.38	0.16	–0.10	0.33	0.32	0.40	0.42	0.53	0.53	24.6%	0.306
Difference	–0.47	–0.70	–0.55	–0.60	–0.57	–0.48	–0.50	–0.45	–0.45	–53.0%	0.083
Mice											
May Die	–0.71	–0.18	–0.34	0.04	0.01	0.23	0.25	0.41	0.40	1.2%	0.370
No Pain	–0.23	0.42	0.14	0.50	0.48	0.57	0.58	0.68	0.67	42.3%	0.293
Difference	–0.48	–0.60	–0.48	–0.46	–0.47	–0.34	–0.33	–0.27	–0.27	–41.1%	0.113
Pain, Illness or surgery	–0.74	–0.25	–0.39	–0.02	–0.03	0.17	0.18	0.33	0.33	–4.7%	0.358
No Pain	–0.23	0.42	0.14	0.50	0.48	0.57	0.58	0.67	0.68	42.3%	0.293
Difference	–0.51	–0.67	–0.53	–0.52	–0.51	–0.40	–0.40	–0.34	–0.35	–47.0%	0.106
Average of Differences (no pain v. may die)	*–0.32*	*–0.47*	*–0.38*	*–0.39*	*–0.39*	*–0.34*	*–0.34*	*–0.29*	*–0.29*	*–35.5%*	*0.058*
Average of Differences (no pain v. pain)	*–0.36*	*–0.55*	*–0.45*	*–0.46*	*–0.43*	*–0.38*	*–0.39*	*–0.35*	*–0.35*	*–41.3%*	*0.067*

Source: MORI/New Scientist

E. To develop a new pain-killing drug
F. To ensure that a new vaccine against the virus that causes AIDS is safe and effective
G. To develop a new vaccine against the virus that causes AIDS
H. To ensure that a new drug to cure leukaemia in children is safe and effective
I. To develop a new drug to cure leukaemia in children.
These are illustrated in Figs 8 and 9.

Exploring the differences across the subgroups, as might be anticipated from the previous analysis, disapproval of the different experiments across the different species and degree of suffering is consistently higher among women, compared with men. Table 12 shows how disapproval varies by gender across a selection of the different types of experiment (the most and least contentious).

Table 12. Men versus women

Experiment	Mice				Monkeys			
%	No pain		May die		No pain		May die	
	M	F	M	F	M	F	M	F
Base:	433	516	433	516	476	584	476	584
A To test whether an ingredient for use in cosmetics will be harmful to people	57	64	81	87	65	71	91	94
B To enable scientists to study the sense of hearing	23	31	52	61	36	44	68	78
C To test whether a garden insecticide will be harmful to people	38	46	61	69	49	57	74	83
H To develop a new drug to cure leukaemia in children	13	17	24	31	20	25	38	48

Source: New Scientist/MORI

The picture among social grades (Table 13) though, is somewhat less clear-cut. Where no pain is involved, disapproval of the selection of experiments among mice is higher among C_2DEs than ABC_1s. Introducing the element of pain, though, has the effect of largely cancelling out the differences—leaving views broadly similar across the social grades.

C_2DEs are more discriminatory in their disapproval of the same experiments where monkeys are concerned than ABC_1s: it makes a difference if the purpose is to cure leukaemia, but otherwise views are broadly comparable with ABC_1s'.

The picture among different age groups (Table 14) is equally 'muddy'. Where mice are used, and no pain is involved, 35–54-year-olds are consistently more likely to disapprove than 15–34-year olds and those aged

Table 13. Social grade

Experiment	Mice				Monkeys			
%	No pain		May die		No pain		May die	
	ABC_1	C_2DE	ABC_1	C_2DE	ABC_1	C_2DE	ABC_1	C_2DE
Base:								
To test whether an ingredient for use in cosmetics will be harmful to people	58	63	85	84	69	67	92	93
To enable scientists to study the sense of hearing	26	32	57	56	37	42	71	75
To test whether a garden insecticide will be harmful to people	39	37	66	64	54	52	78	80
To develop a new drug to cure leukaemia in children	12	18	25	30	18	27	37	48

Source: New Scientist/MORI

55+, this generally follows through when the mice may die, with the exception of curing leukaemia, where the differences are much flatter (but much lower overall).

With the exception of garden insecticides, views are more consistent across the age groups for monkeys than they are for mice—both if death is involved and if it is not.

Table 14. Age

Experiment	Mice						Monkeys					
%	No pain			May die			No pain			May die		
	15–34	35–54	55+	15–34	35–54	55+	15–34	35–54	55+	15–34	35–54	55+
To test whether an ingredient for use in cosmetics will be harmful to people	58	68	55	87	85	81	70	68	67	93	94	90
To enable scientists to study the sense of hearing	26	32	33	57	62	51	43	40	37	76	75	68
To test whether a garden insecticide will be harmful to people	40	49	37	68	68	58	59	53	47	82	80	75
To ensure a new drug to cure leukaemia in children	11	19	16	24	29	31	21	22	25	44	41	44

Source: New Scientist/MORI

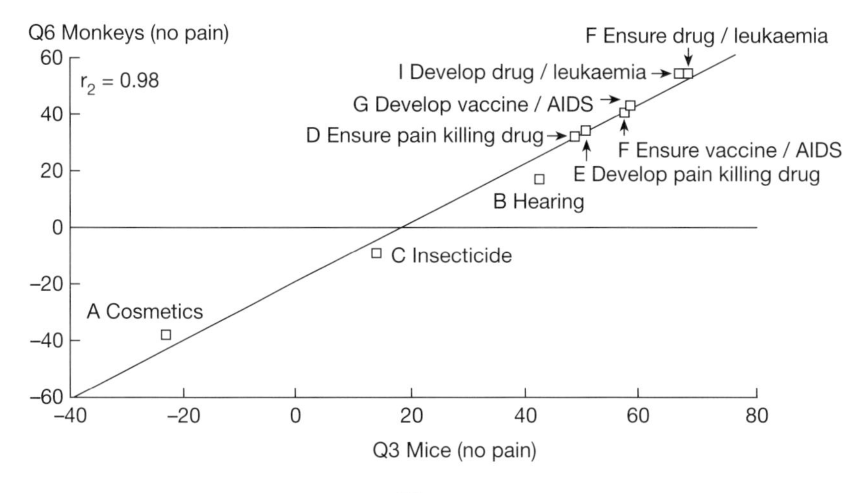

Fig. 8.

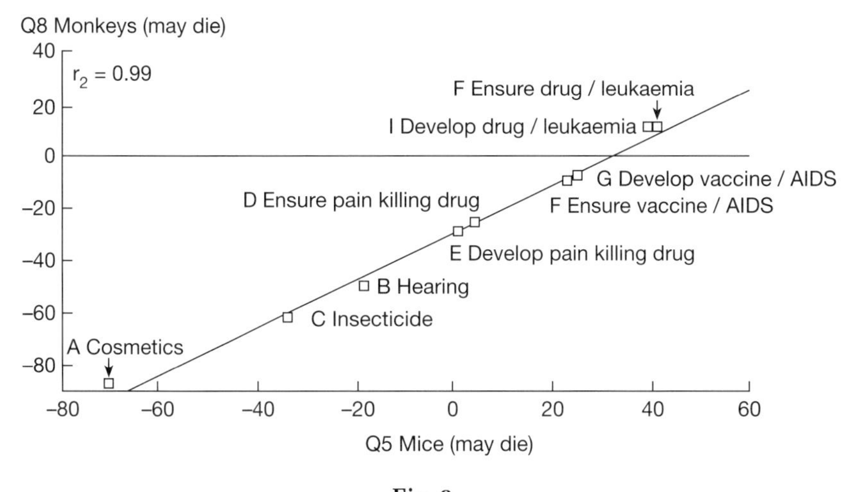

Fig. 9.

As indicated above, the survey was used to determine the incidence and profile of the British public's 'scientific behaviour', from taking part in demonstrations against animal experimentation to visiting the countryside, with the parallel opportunity to cross-tabulate the behavioural data with the attitudinal data on animal experimentation. Table 15 replicates the results.

Reflecting earlier sub-group discussion, disapproval of experiment H (Table 16) is higher among vegetarians, members of animal welfare organizations, and those who have signed a petition on an animal welfare issue.

Table 15.

	SHOWCARD Which, if any, of the following things would you say you have done over the last two years or so? Just read out the letter or letters that best apply. MULTICODE OK	%
A	Avoided genetically-modified foods	20
B	Been a vegetarian/vegan	8
C	Been a member of an organisation involved in animal welfare	7
D	Been horse-riding	8
E	Been to a farm	35
F	Been to a laboratory where experiments are conducted on animals	1
G	Been to a zoo/safari park/animal park	33
H	Been fishing	11
I	Bought 'cruelty-free' cosmetics, not tested on animals	32
J	Gone for a drive in the countryside/along the coast	69
K	Owned a pet	55
L	Taken part in a protest or demonstration against a blood sport – e.g. fox hunting, hare coursing, beagling	2
M	Taken part in a protest or demonstration against experiments on animals	1
N	Taken part in a protest or demonstration against the shipment of live farm animals	1
O	Read a science magazine	22
P	Signed a petition on any animal welfare issue	20
Q	Studied a scientific subject	12
R	Taken part in any blood sport – e.g. fox hunting, hare coursing, beagling	1
S	Walked in the countryside/along the coast	70
T	Worn a fur coat	1
U	Written a letter to an MP or editor of a newspaper/magazine protesting against animal experiments	3
V	Written a letter to an MP or editor of a newspaper/magazine protesting against any blood sport – e.g. fox hunting, hare coursing, beagling	2
W	Written a letter to an MP or editor of a newspaper/magazine protesting against the shipment of live farm animals	2
X	You or a close family member have taken a drug prescribed by a doctor for a serious illness	34
Y	You or a close family member have taken a drug prescribed by a doctor for a serious illness that you knew had been tested on animals	6
	Other	*
	None of these	7
	Don't know	0

Base; All respondents (2009: full sample)

Table 16. Lifestyle groups

H: Leukaemia	Tot.	A	B	C	D	E	G	H	I	J	K	O	P	Q	S	X	Y	*
Mice																		
Base:	949	209	80	79	77	331	323	106	319	683	549	221	208	118	688	349	65	26
No Pain	15	16	26	26	9	13	14	16	19	14	16	13	22	10	14	12	16	
Death	28	31	37	40	20	25	27	31	33	27	31	27	37	20	26	25	26	
Monkeys																		
Base:	1060	208	80	70	72	359	352	115	340	714	577	202	205	104	713	340	60	26
No Pain	23	23	30	39	16	17	20	18	21	19	23	22	31	19	20	20	13	
Death	43	42	52	62	29	35	42	35	46	39	46	36	55	34	42	39	21	

* Combined category: 'taken part in a blood sport' or 'worn a fur coat'.
Source: New Scientist/MORI

For an experiment at the other end of the spectrum (Table 17), these groups are joined by those who have avoided genetically modified food or read a science magazine.

Follow-on work on animal experimentation for the MRC

The main objective of the follow-on research for the Medical Research Council was to examine the general public's views on animal experimentation, to inform MRC communications work. The research was designed to address the following questions:

- Current level of public knowledge about animals in research.
- The assumptions and attitudes held by the general public about animals in research, e.g. is this perceived to be ethical? cruel?
- What values influence attitudes and assumptions? What sources of information influence opinion?
- How different people respond to different arguments and information.
- Current concerns about animal experimentation.
- Who, if anyone, the general public trusts with respect to animal experimentation.
- Awareness of any controls on animal experimentation in the UK and internationally.
- How opinion shifts, if at all, when people are well informed.
- What questions the general public want answered.

In total, 1014 interviews were carried out among adults aged 15+, in Great Britain, between 1 and 26 September 1999.

Level of interest in animal experimentation

More people say they have (at some time in their lives) discussed the issue of animal experimentation with another person, than have not: 53% have not, and 11% can't remember (Table 18). Over a quarter say they have discussed it in the previous three months, which seems high. There were quite a few media stories about it from June to August, prior to fieldwork, which would help explain this seemingly high figure. (The Hillgrove Farm cat laboratory closed down just prior to main-stage fieldwork, which received TV and press coverage; there was also some coverage on excess breeding in the broadsheets just prior to main-stage fieldwork; and on 7 June, Channel 4's *Equinox* about phantom pain gave detail of the Silver Spring macaque monkey experiments. In all, eight different animal experimentation stories are known to have run during this period.) In addition the 'recency effect' could be playing a part (people feeling that events took place more recently than they actually did). MORI has observed this in previous behavioural data, e.g. on when

Table 17. Lifestyle groups—experiment A

A: Cosmetics Leukaemia	Tot.	A	B	C	D	E	G	H	I	J	K	O	P	Q	S	X	Y	*
Mice																		
Base:	949	209	80	79	77	331	323	106	319	683	549	221	208	118	688	349	65	26
No Pain	61	70	69	74	65	59	59	66	69	61	63	58	75	57	61	60	58	50
Death	84	89	91	98	83	83	86	84	93	86	88	84	95	86	86	85	80	50
Monkeys																		
Base:	1060	208	80	70	72	359	352	115	340	714	577	202	205	104	713	340	60	26
No Pain	68	74	75	83	60	68	71	60	71	69	68	69	82	76	69	64	64	50
Death	92	97	98	99	93	94	93	93	96	94	94	97	98	95	94	93	94	86

Source: New Scientist/MORI

Table 18. Q. When, if at all, did you last discuss the issue of animal experimentation with another person such as a close relative, friend or colleague?

	Cumulative total (%)
In the last week	8
Up to 1 month ago	17
Up to 3 months ago	28
Up to 6 months ago	35
Up to 1 year ago	41
Up to 3 years ago	47
Ever	53
Never	36
Don't know/can't remember	11

Source: MORI/MRC
Base: All (1014)

last visited the optometrist or dentist (where practice figures suggest that actual visits perhaps took place longer ago than reported—at least before payment was introduced).

Most people who have discussed animal experimentation with someone were prompted to do so by hearing something on the news or seeing a television documentary; over a quarter mention each spontaneously, while 'Read an article in the press' is mentioned by fewer (15%). TV/broadcast news is probably the most likely trigger for people, particularly given the strong visual images which animal experimentation can evoke. However, in MORI's experience television tends to be over-reported as a source of information (for example, some people report having seen an advertising campaign on television, when it only appeared in the press). While broadsheet readers display a marginally greater likelihood to have discussed animal experimentation as a result of a newspaper article, this difference is not statistically significant. (A difference of 12 points or more would be required on these size samples; the actual difference is just five points.)

We therefore find the level of interest displayed in the quantitative phase largely consistent with the focus groups. While 67% say they are 'interested' (to discuss the topic or hear more about it), only 14% say they are 'very interested' (Table 19). In MORI's experience, this is a little on the low side for a 'very' proportion.

Also, one in eight people in the quantitative research say 'It does not bother me if animals are used in experimentation' and/or 'I am not interested in the issue of animal experimentation'. This translates into over 5 million adults in Great Britain (based on an adult population of 44 million adults aged 15+ in Great Britain) who say the former or who say the latter, and over 8 million who say one or other or both.

Table 19. Q5. How would you rate your interest in the issue of animal experimentation?

	Total (%)
Very interested	14
Fairly interested	53
Not very interested	24
Not at all interested	8
Don't know	1
Interested	67
Not interested	32

Source: MORI/MRC
Base: All (1014)

Unnecessary duplication in animal experimentation, on the part of industry, was cited in all four group discussions. Pharmaceutical companies and then household products companies attract greatest criticism for this; they are said to be competing to develop new medicines and produce new cleaners, respectively. Cosmetics companies are also mentioned in this respect, even though some knew that cosmetics testing on live animals no longer takes place in the UK.

> *There's quite a lot of research, you hear, done by universities, medical institutions, that are profit-based ... I've never heard when a company comes to making a drug, selling it at a loss.*
>
> Man, 45+, ABC_1, Leeds

Pooling of results and research evidence should be essential, according to some focus group participants—a theme mentioned in all four groups. Communication between companies would reduce duplication and mean that fewer animals would need to be experimented on.

> *If drug companies are researching the same product, the cure for cancer is the Holy Grail, isn't it? Well, they're all at it. It's a race, so they'll all be testing and almost certainly all be moving along the same lines. So, a lot of work must be replicated and repeated.*
>
> Man, ABC_1, Northwood, Middlesex

The media (largely through TV advertising) is considered to be encouraging people to buy household and other products, and many people admitted to being susceptible to advertising.

There was therefore a suggestion of a 'corporate machine'—with companies competing to develop products, the media advertising them, and the consumer being convinced that they need to buy them. Some talked of the many products (particularly household cleaners) which they had bought and wondered for a moment just how much animal experimentation had been conducted to produce them. One older man in Leeds, turning to other participants in the group, said they may be worried about animal experimentation in the focus group but they would still be

Table 20. Q2. On this card is a list of situations for which animal experimentation might be carried out. Could you read through the list and tell me which, if any, of these situations you think animal experimentation is always justified, sometimes justified or never justified?

	Always justified (%)	Sometimes justified (%)	Never justified (%)
Medical:			
Life-threatening diseases such as cancers	42	35	19
Life-threatening diseases such as AIDS	33	38	25
Ways of preventing diseases, e.g. vaccines	29	43	24
Improving medical treatments and surgical techniques	23	39	33
Testing potential new medicines	21	45	30
To learn how cells work	13	43	34
Treatments to improve quality of life, e.g. HRT	11	35	46
Agricultural and veterinary:			
Researching animal diseases	31	48	16
Improving livestock welfare, e.g. preventing disease in cattle herds or preventing stress in transported animals	24	43	27
Improving livestock to make sheep woollier and meat leaner	4	17	73
Safety testing:			
Testing chemicals in the workplace	8	25	61
Testing the safety of household products, e.g. disinfectants, DIY products	4	15	77
Testing the safety of cosmetics, e.g. skin care products, make-up	4	9	85

Source: MORI/MRC
Base: All (1014)

'conned' by the advertisers into buying products. (He himself was 'not bothered' about animal experimentation.):

> *I was looking only the other day. I've got four or five things under my sink; they all do the same thing ... And this is supposed to do a little bit more than this one, but the three out of four come from the same company, so what difference can there be?*
>
> Woman, 54+, C_2DE, Northwood, Middlesex

This respondent talked of the financial loss which a company could face if it did not carry out comprehensive testing on cosmetics, prior to launch:

> *If you're going to put something on your face that hasn't been tested on some kind of skin, then when you're talking about multi-million pound companies that are maybe bringing an awful lot of money in to a product, if that goes on to a surface and it's going to have a bad reaction, then they're going to lose an awful lot of money.*
>
> Man, 20–35, ABC_1, Northwood, Middlesex

Consistent with the focus groups, few in the quantitative survey feel that animal experimentation is always justified for testing chemicals in the workplace (though a distinction was drawn between chemicals which could contravene health and safety regulations or harm workers, and other chemicals (which could account for the higher 'sometimes justified' figure), or for testing the safety of household products and testing the safety of cosmetics. However, 4% (rather than no one) said that cosmetics testing is always justified and 9% that it sometimes is. In the focus groups, people did distinguish between cosmetics 'that supposedly make you look good' and 'those that mask disfigurements'. The latter are classed as being 'medical' and therefore animal experimentation for this purpose is sanctioned:

> *I mean, surely they should have enough knowledge now not to have to test lipstick on a rat or something like that?*
>
> Woman, 45+, ABC_1, Leeds

MORI's work for the *New Scientist* examined views towards basic medical research, like enabling scientists to study how the sense of hearing works; 70% in that poll approved of it, in the case where mice would not be subjected to pain. This was then explored further in the focus groups for the MRC. Some participants did not have a view. However, research on the ear to understand hearing was thought unnecessary by some of the younger people as they think that animals have different ears and cannot tell the researcher how loud a noise is to them. Testing on humans, dead or alive, for this type of research was cited as the preferred option in the focus groups. In the focus groups, 'Household cleaners', are put closer to 'Cosmetics' than to 'Medical purposes' on respondents' continuum of the degree of acceptability. They largely oppose animal experimentation to test (new) household cleaners.

Other categories were difficult for participants to place (even household cleaners were difficult because they could see a use for them, and most feel that safety of people is paramount, but they expressed concern about animal welfare).

Nevertheless, for both agricultural and household products, most agree that if there is any risk involved to human health, these situations need to be adequately tested, even if this means using animals. This applies to any chemicals used in agriculture which could enter the human food chain. An example given by participants of a problem caused from a lack of testing is the use in roofing of asbestos, now known to cause asbestosis.

The main reason, in the group discussions, why people think animal experimentation takes place is that most people feel there are no, or limited, alternatives. Alternatives put forward by participants include: human volunteers, computer modelling, cell/tissue/skin cultures in test tubes, experiments on organs, and homeopathic methods. Cloning of animals is mentioned (by some) for the purpose of animal experimentation. However, we did not get the impression that cloning of animals was widely supported for this purpose. Over nine in ten, in the quantitative survey, feel that 'There needs to be more research into alternatives to animal experimentation'.

While a majority in the group cannot see any alternatives for most of the current research that they are aware of, e.g. 'spraying chemicals in eyes' and 'cures for diseases', they would like to see more investment into alternatives. Some people feel that alternatives might be more expensive and therefore not readily pursued. Animals are described as being a cheap source which reproduces frequently, but with some disappointment on the part of the respondents:

> *There's no alternative really. A human's not going to go to somewhere and have stuff squirted in his eyes ...*
>
> Man, 25–44, C_2DE, Leeds

Our qualitative findings agree that most people feel that animal experiments will always be conducted, but there is also a strong feeling that even if the animal suffers during tests they should be kept in comfortable conditions so that they do not suffer unnecessarily. After experimentation, the animals are not thought to make good pets, but if they are destroyed, it should be done humanely.

Animal suffering is also most frequently cited on the subsequent prompted question as a factor that should be taken into account in the current regulatory system regarding animals in experiments for medical research purposes (56%). However, perhaps because it was a prompted question, many other factors are cited too (Table 21).

Table 21. Q24. Using this card, how strongly do you agree or disagree with these more general statements about animal experimentation?

		Agree (%)	Disagree (%)
B	There needs to be more research into alternatives to animal experimentation	91	3
F	Animal experimentation will always be used for research purposes	72	15
C	I can accept animal experimentation so long as there is no unnecessary suffering to the animals	69	21
D	I would like to know more about animal experimentation before forming a firm opinion	65	18
A	I can accept animal experimentation so long as it is for medical research purposes	64	24
L	I agree with animal experimentation for all types of medical research, where there is no alternative	61	26
J	Animal experimentation for medical research purposes should only be conducted for life-threatening diseases	58	27
I	Animal experiments for medical research purposes are a necessary evil	54	28
E	I do not support the use of animals in any experimentation because of the importance I place on animal welfare	39	37
M	I agree with animal experimentation for all types of research where there is no alternative	27	60
K	The Government should ban all experiments on animals for any form of research	26	55
G	It does not bother me if animals are used in experimentation	12	79
H	I am not interested in the issue of animal experimentation	12	75

Analysis of all the questions has been carried out by the following combinations of these statements:

	(%)
Conditional acceptors—agree with A, C, J, or L	84
Unconditional acceptors—agree with G or M	32
Disagree with animal experimentation—agree with E or K	44

Source: MORI/MRC
Base: All (1014)

As mentioned, the emphasis of discussion in the groups and the quantitative survey, is on animal welfare during experimentation but also on the laboratory living conditions for the animals. One in ten think more inspections are needed and around one in six would like to see random spot checks of all regulated activities (Table 22).

One of the objectives of this research was to determine whether the public perceives animal experimentation as being ethical. This is a difficult question to answer. The majority of people in the groups regard it as a 'necessary evil', 'conundrum' or 'dilemma', and we know that two-thirds can accept it so long as there is no unnecessary suffering to animals. Our assessment would therefore be that the public does not allow themselves to think of animal experimentation along the lines of its being ethical or unethical. Rather, they think about whether it *has* to happen (because there is no alternative) or whether an alternative could be found (something of great importance to them).

As mentioned earlier, the vast majority agree that they 'wouldn't be surprised if some animal experiments go on behind closed doors without an official licence'. In the groups, there was fairly frequent mention that the current regulatory system may be corrupt, with mention of 'backhanders' to government to turn a blind eye. However, this did not seem to based on any actual knowledge or experience; by their own admission

Table 22. Q10. Thinking now about the use of animals in experiments for medical research purposes, what factors, if any, would you take into account if you were deciding whether such experiments were right or wrong? (UNPROMPTED QUESTION)

	Combined mentions (%)
The suffering/pain the animal might endure/if no unnecessary suffering	33
The importance of research to human health	20
Do research if it is for a life-saving cure/treatment	20
Only if no alternative	16
If it is for long-term or chronic illnesses, e.g. diabetes, arthritis or Parkinson's disease	9
The animal welfare within the laboratories	9
Whether it was for the general good	9
Whether it is absolutely necessary to have the knowledge that the experiment will provide	9
Aim of development/Why are they doing it	8
Not allow research for cosmetics, such as lipsticks and mascara	7

Source: MORI/MRC
Base: All (1014)

(from the focus groups and the quantitative survey), people know very little about the regulatory system (Table 23).

Only four in ten say Britain probably has tough rules and regulations on animal experimentation, but a further three in ten are neutral or express no opinion. As one might expect, those who trust the regulatory system are far more likely to say the rules are tough than those who do not (67% versus 31%). Communicating with the public about the current system may increase their likelihood of saying the rules are tough, but some people may still feel that they are not tough enough or that experimentation should not take place at all. 'Actives', for example, are significantly less likely to say that Britain has tough rules—and they are a group whose views would be less likely to change even if they were given information about the regulatory system in place.

Fewer than three in ten say that they expect that the rules in Britain are well enforced and here over four in ten actively disagree with the statement, suggesting some real concern about enforcement. This was also expressed in the focus groups: some participants talked of the need for spot checks on laboratories. One said that when the inspectors come, they would probably be ushered to the lab that was behaving itself, and away from the one that was breaching the rules. This concern about enforcement was also voiced in the qualitative workshops on regulation of biotechnology in MORI's work for OST, and is linked to who people trust (discussed later) as well as their concern about companies' profit motive.

In the groups for MRC, a permanent inspector is put forward as a solution to the problem of breaching the rules. One person who knew more than most about scientific experiments said they felt inspections *were* conducted to a very high standard, but Hillgrove Farm was cited by others as evidence of lax inspections.

This quote from one woman suggests greater faith in government than corporations, though people generally expressed little faith in either:

> *I'd like to see someone who worked in the laboratory whose sole responsibility was for the welfare of the animals, who wasn't answerable to the company but answerable to the government.*
>
> Woman, 45+, ABC_1, Leeds

Two-thirds say they have a lack of trust in the regulatory system, and they are much more likely to be 'Actives' (78%). The vast majority (83%) feel that there may be unnecessary duplication of experiments (mentioned earlier in this report). Even among groups who are largely supportive of animal experimentation (e.g. men and ABs), most say that they feel that needless duplication is taking place.

As well as a concern for animal welfare, the benefit of the research to human health is a major consideration in the qualitative research and over two in five suggest it as an important factor that should be taken into account in regulation.

Table 23. Q23. Using this card, how strongly do you agree or disagree with the following statements about the rules and regulations governing animal experimentation?

	Agree (%)	Disagree (%)
I wouldn't be surprised if some animals experiments go on behind closed door without an official licence	89	4
I don't know a lot about regulation regarding animal experimentation	86	5
I feel that unnecessary duplication of animal experiments may go on	83	5
I have a lack of trust in the regulatory system about animal experimentation	65	11
Britain probably has tough rules governing animal experimentation	42	28
I expect that the rules in Britain on animal experimentation are well enforced	29	41

Source: MORI/MRC
Base: All (1014)

One in ten (at Q11—Table 24) do not put forward any factors which should be taken into account in the current regulatory system regarding animal experiments for medical research purposes, on the grounds that they feel such experiments are wrong (at Q11Z). As one might expect, this proportion increases among the group who say that they disagree with animal experimentation (Q24—Table 21), but only to 22%, and with fewer than half a percent age point of this group replying 'Don't know' at this question. This means that the majority of this group did cite factors which should be taken into account in the current regulatory system. For example, 32% of this group said 'Do research if it is for a life-saving cure'.

A quarter of each of the following groups at Q11 do not put forward any factors, on the grounds that they feel such experiments are wrong: 'Super-actives' (23%), those very interested in animal experimentation (25%), and those who do not think that medical research is ever justified (27%). One in five of those aged between 45 and 54 hold this view.

However, 55% disagree that the government should ban all experiments on animals for any form of research (at Q24K), a quarter disagreeing strongly, and disagreement increases considerably to 77% among the group who say experiments for medical research purposes are 'always justified'. Though lower than average, even 47% of 'Actives' and 43% of 'Super-actives' disagree with this statement, and 65% of conditional acceptors disagree (i.e. a significantly higher proportion than average).

Table 24. Q11. Which, if any, of the following factors do you think should be taken into account in the current regulatory system regarding animals in experiments for medical purposes?

		Animal experimentation justified for medical research			Animal experimentation			Trust in the rules and regulations	
Base:	Total (1014)	Always[1] (380)	Sometimes[2] (550)	Never[3] (320)	Conditional acceptors (845)	Unconditional acceptors (318)	Disagree with it (444)	A great deal/ a fair amount (240)	Not very much/ none at all (579)
Top ten answers	(%)	(%)	(%)	(%)	(%)	(%)	(%)	(%)	(%)
D The suffering/pain the animal might endure/if no unnecessary suffering	56	61	62	43	61	54	46	59	57
Y Do research if it is for a life-saving cure/ treatment	51	72	57	24	59	62	32	68	44
W Not allow research for cosmetics, such as lipstick and mascara	47	46	54	41	51	41	42	48	48
L Whether the experiment is stopped as soon as the animal feels pain	45	46	54	34	50	42	40	42	48
laboratories	43	49	50	32	48	41	32	48	41

Table 24. *Continued*

			Animal experimentation justified for medical research		Animal experimentation			Trust in the rules and regulations	
Base: Top ten answers	Total (1014) (%)	Always[1] (380) (%)	Sometimes[2] (550) (%)	Never[3] (320) (%)	Conditional acceptors (845) (%)	Unconditional acceptors (318) (%)	Disagree with it (444) (%)	A great deal/ a fair amount (240) (%)	Not very much/ none at all (579) (%)
P Whether it was well supervised to ensure high standards of animal welfare	44	52	50	29	49	45	31	56	39
B Only if no alternative	43	44	53	31	49	40	32	50	42
E The animal welfare within the laboratories	43	49	50	32	48	41	32	48	41
G The importance of research to human health	43	62	47	17	50	53	23	60	33
X If it is for long-term or chronic illnesses, e.g. diabetes, arthritis or Parkinson's disease	43	63	48	15	50	54	24	61	35

Table 24. *Continued*

		Animal experimentation justified for medical research			Animal experimentation			Trust in the rules and regulations	
Base: Top ten answers	Total (1014) (%)	Always[1] (380) (%)	Sometimes[2] (550) (%)	Never[3] (320) (%)	Conditional acceptors (845) (%)	Unconditional acceptors (318) (%)	Disagree with it (444) (%)	A great deal/ a fair amount (240) (%)	Not very much/ none at all (579) (%)
F Whether spot checks on laboratories were carried out	40	47	45	27	45	40	27	44	38

Source: MORI/MRC

[1] Answered at least two 'Always justified' from Q2 E, F, H, I, J (Table 20)

[2] Answered at least two 'Sometimes justified' from Q2 E, F, H, I, J

[3] Answered at least two 'Never justified' from Q2 E, F, H, I, J

Although there is some confusion in the groups over who governs regulation, the general consensus of opinion is that 'government' are the regulators.

One participant said that different ministries had different sets of regulations and others agreed that this should be the case if it was not already. One or two people did mention 'licences' but gave no more detail.

It is suggested that an 'independent regulatory body' should govern regulation but some people say that no one is independent, a point also mentioned by workshop participants in MORI's work for OST.

> *A cross-party type ombudsman or inspectorate made up of partly research company representatives, people from the industry ... partly from government, partly from the animal welfare groups or any other funded relevant groups.*
>
> Man, 20–35, ABC_1, Northwood, Middlesex

It is felt that anyone involved in animal experimentation would be biased towards lax regulations but the difficulties are recognized in finding a completely neutral person in the field of science with an awareness of the issues. Again, a communications campaign should convey precisely who is involved in the regulatory system and the safeguards undertaken to ensure that standards set are adhered to.

The idea of self-policing was rejected, as a voluntary code is not thought to work.

There was also a view, mentioned by some in the focus groups, that over-regulation on animal experimentation could be a problem as it could stifle medical developments and result in additional costs to the consumer.

One participant mentioned that human drugs used to be used on animals but that there are now new regulations that state that all the drugs need to be re-licensed so that they are specific for animals. This involves excessive animal experimentation and the drugs (she felt) are now sold at much higher prices. She suspects it was a money-making exercise.

According to some participants in the groups, stringent regulation in the UK could also simply result in experimentation being conducted abroad, e.g. in Europe or America, where there are thought to be more relaxed regulations.

Trust

Far fewer people say they trust the current rules and regulations governing animal experimentation than do not (24% Great deal/Fair amount, 57% Not very much/None at all). Yet by their own admission, few people say they know anything about the rules and regulations—and that was evident from the focus groups. This is much more likely, then, to reflect a more deep-seated feeling of lack of trust in policy-making institutions. 'Actives' are significantly less likely to place their trust in the rules and regulations, but conditional acceptors are a little more likely to do so.

Conditional acceptors, to remind the reader, are those who agree with one or more of the following: they can accept animal experimentation so long as it is for medical purposes, so long as there is no unnecessary suffering to animals, so long as it is only conducted for life-threatening diseases, and/or so long as there is no alternative. Because they place one or more of these conditions on their acceptance of animal experimentation, it is perhaps surprising that they are more likely than average to trust the regulatory system. Even so, conditional acceptors who do not trust the regulatory system far outweigh those who do: 28% trust it, but 51% do not. Given that conditional acceptors represent such a large proportion of the public (84%), their feeling of unease with the regulatory system is particularly important. We saw earlier that the public has concerns about enforcement of regulation which could explain some of this unease. The reasons for the responses given at this question are discussed further below.

Those who gave a correct or near-correct response about the proportion of medical research which involves animal experimentation are significantly more likely to place their trust in the regulatory system but, again, more of them do not trust the regulatory system, than do (39% Great deal/Fair amount of trust, 49% Not very much/None at all). The relative proportions for those who say the current rules are tough are similar, at 38% and 42%. As mentioned earlier, informing people about the current system may increase trust in some cases, but other groups may remain sceptical.

When those saying they have 'Not very much trust' or 'No trust at all' in the regulatory system are asked why, almost three in ten say because they do not have enough information, and one in six because they do not know enough. So, informing people could reassure some members of the public but it is interesting that when so many people know so little, they opt for a negative rating on trust, rather than a neutral one or no opinion. Trust in governments to provide honest and balanced information about the regulatory system receives a poor 'Net trust' rating of –43 (6% trust governments, but 49% do not), which helps explain why the trust figure for regulation of animal experiments is poor.

At the follow-up question on reasons why people did not trust the current regulatory system, one in four of this group (the second highest mention) said 'Because I do not trust those who regulate; and around one in ten 'Because it's cruel' or/and 'Because they are only interested in profits' or/and 'The system is corrupt, or/and 'More inspections are needed' or/and 'Because we don't see the results of regulation' or/and 'Because the regulation does not seem to have an impact on what we are doing'.

The net trust figure for governments in MORI's work for OST was less unfavourable on a similar question ('Which, if any, of the following types of people or institutions would you trust to provide you with honest and balanced information about biological developments and their regulation?') 'Trust' stood at 19%, 'Not trust' at 39% and so net trust was –20 (Table 25).

There are many possible reasons for the low net trust rating on this survey:

- the list size (though this was similar, at 20 or OST and 22 for MRC);
- the categories on the list: OST's list included farmers, retailers and sociologists—none of which received good ratings; MRC's list included celebrities, charities researching heart disease, cancer, etc., anti-vivisection groups and the Medical Research Council (Both the MRC and charities receive positive net trust ratings, which could have acted to pull down the rating for government somewhat; the MRC's rating is very likely to have been boosted by its name—we suspect that few people would know what it actually does, beyond guessing from its name.);
- the actual topic—animal welfare groups and vets appeared on both lists and receive a very high net trust rating on the MRC's survey (+50 for vets and +36 for animal welfare groups), no doubt because of their relevance to the topic and people's concerns about animal welfare; for the OST study, net trust for vets was +12 and for animal welfare groups it was –19;
- the position of this question in the questionnaire and what had preceded it (Q24/25 for OST; Q16/17 for MRC. However, there is no real evidence to suggest that the MRC's questionnaire had more statements about government prior to these questions being asked);
- respondents' knowledge about the topic and (therefore) the degree to which feelings could run high about it. Certainly people seem to know more about animal experimentation than they do about any aspect of biotechnology and this could have therefore made them even more critical of regulatory authorities;
- the media coverage prior to the survey. However, if anything, this would have been far more likely to have made respondents critical of government in the MORI survey for OST than in this study for the MRC. Fieldwork for the OST's quantitative project followed the intense media coverage on GM food, which began in February and continued daily until after fieldwork ended on 4 April 2000;
- the fieldwork period (March/April for OST, September for MRC). There is no evidence at all, however, from other MORI research that

Table 25. Q16. Which, if any, of the following types of people or institutions would you trust to provide you with honest and balanced information about animal experimentation?
Q17. And which, if any, of the following types of people or institutions would you not trust to provide you with honest and balanced information about animal experimentation?

	Q16 Trust (%)	Q17 Not trust (%)	Net trust (±%)
Animal welfare/environmental/anti-vivisection groups (COMBINED)	61	36	+25
Vets	56	6	+50
Animal welfare groups	54	18	+36
Advisory body	33	27	+6
The Medical Research Council	32	18	+14
Charities researching diseases, e.g. heart disease, cancer	27	12	+15
An advisory body to Government, composed of people representing different viewpoints	25	15	+10
Environmental groups	25	16	+9
Doctors/pharmacists/chemists	21	30	–9
Teachers/universities	20	21	–2*
An advisory body to Government, composed of experts	20	24	–4
Scientists	20	36	–16
Anti-vivisection campaign groups	19	28	–9
GPs/Family doctors	18	13	+5
Universities	16	17	–1
Religious organisations	10	23	–1423
Teachers	8	14	–6
The General Public	8	22	–1323
Consumer groups	8	22	–14
Celebrities/well-known personalities	8	28	–2123
The Media	8	44	–3723
Pharmacists/Chemists	7	26	–2023
Patients	6	12	–723
Governments	6	49	–43
Industry/manufacturers/ pharmaceutical companies combined	5	66	–6223
Pharmaceutical companies	4	54	–50
Industry/manufacturers	1	49	–4723
None of these	6	1	+423

Source: MORI/MRC
Base: All (1014)
* The 'Net trust' figures have been calculated from the numbers of respondents, not percentages.
Hence the figure differs by one point from a straight subtraction.

> satisfaction with the government (generally) fell between March and September. The proportion satisfied/dissatisfied with 'the way the government is running the country' was 47%/41% in March and 47%/40% in September (MORI/*The Times*).

We feel the most likely explanation is that groups that are either perceived to be more relevant to provide information on animal experimentation (e.g. animal welfare groups, and medical charities, and, less so, environmental groups) or/and who sound relevant (the Medical Research Council) are effectively pulling down rating for 'governments'. Also, it is possible that governments may not seem to be as relevant to provide information about animal experimentation, as biotechnology.

There is more trust in an advisory body reporting to Government, than governments directly. One in five trust an advisory body to Government composed of experts and this rises to a quarter if the body is composed of people representing different viewpoints. The 'net trust' figures are –4 and +10 respectively.

Groups which receive poor or fairly poor net trust ratings (to provide honest and balanced information) are: pharmaceutical companies (–50), industry/manufacturers (–47), governments (mentioned above, –43), the media (–37), celebrities/well-known personalities (–21), pharmacists/chemists (–20), scientists (–16), consumer groups (–14), religious organizations (–14), and the general public (–13).

Feelings towards industry, and pharmaceutical companies specifically, have been discussed in earlier sections of this report. Concern about them stems from a feeling of unnecessary duplication of animal experiments, as well as the profit motive, cited in the focus groups as being behind everything.

Retailers too (not measured) specifically in the quantitative phase) were generally seen in the focus groups as being untrustworthy. Products sold in shops, e.g. Body Shop, say they are 'Not tested on animals', but, according to one man in Morley 'the components would have been tested at some point'. This led to a debate about whether manufacturers' claims are correct and whether any 'animal testing of the products or components' should be written on the label. Although some would like any experiments conducted on animals to be listed next to the ingredients of a product, others find the idea very unsavoury or unworkable (i.e. no space on the label) and do not see how it would help anyway.

The poor overall rating for well-known personalities could partly be explained by the fact that they do not appeal to everyone. This point was made by focus group participants.

Another reason for the low rating is that well-known celebrities do not appeal to everyone. ABs and men give them poor ratings (–29 and –24 respectively). There is little difference by age (the young do not rate personalities any more favourably, for example), though older people aged

65+ do have a net trust rating of –7. However, they are less inclined to express an opinion at either of the trust questions.

The negative ratings for industry and the media in this study for the MRC are almost identical to those found for the OST on 'biological developments and their regulation'. However, the ratings on trusting scientists and pharmacists/chemists, like governments, are more negative in this work for the MRC.

Implications

Public attitudes towards animal experimentation are far from clear-cut. While, the initial 'knee-jerk' reaction may be one of substantial opposition, the public is receptive to messages explaining (justifying) what benefits such experiments may bring. One in four of the British public is implacably opposed to any animal testing, and only 4% of the public are strongly in favour, thus leaving seven in ten who are open to persuasion, and who, as has been shown, make considered judgements based on not only their own prejudices, but several other important factors as well.

When exploring the factors that drive opinion, three can be seen to play a pivotal role in the public's acceptance, or not, of a given experiment. The key driving factor is the purpose of the experiment—those with a clearly defined, medical aim are viewed very differently from, particularly, cosmetics testing. Second is the extent to which the animal suffers pain or illness, or undergo surgery, or may die as a result of the experimentation. And, third, is the species to be experimented on—though this is seemingly less important than the previous two, at least as between mice and monkeys.

However, the public is relatively unconcerned with the issue of how close the experiment is to actually being used on humans (i.e. developing drugs, rather than testing efficacy) and does not distinguish significantly between the two.

In many ways, based particularly on their consistency and 'logic' of answering, the public overall, seem to make a careful evaluation of the facts presented before arriving at an 'approve' or 'disapprove' decision. However, while this follows for the overall findings, with the exception of gender and perhaps some lifestyle groups which suggest a greater 'respect' for the environment, it is very difficult to isolate clear and consistent reactions among the different subgroups to all the dimensions and scenarios presented. This suggests to us that, for the most part, the reaction is very much an individual one—some of us react more to the experiment's purpose, others to the thought of the animal suffering pain or dying, and others to the species, tending to support the hypothesis that the decision-making process is more value- than demographically driven.

What, however, is much more straightforward to pin down is the effect that an explanation can have on the public's acceptance of animal experimentation—the greater the understanding, the greater the acceptance. Nevertheless, we would suggest that no amount of explaining of some experiments, on some species, and where the effect on the animal is likely to be severe, will ever gain any degree of credence with the British public.

Conclusions

This 'flashpoint' issue of animal experimentation well illustrates the problem of 'public understanding of science' which captured attention of the Lords' enquiry. As they stated it, but not all would agree, 'there is a new humility of the part of science in the face of public attitudes'. Sir Walter Bodmer (2000) has been in the forefront of the 'public understanding of science' movement for more than a quarter of a century, but still observes: 'it is clear that simply explaining to the public out there is not enough. There must be a two-way dialogue and an understanding by the scientists of the public as well as by the public of science.' The appointment of Jonathan Porritt to head the government's task-force on sustainable development is a good and brave start, indicating that the Department of the Enviroment, Transport and the Regions and No. 10 are on board on beginning the dialogue.

Science is under attack, and probably feels hard-done-by, but is not alone. Business, government, the universities, charities and other NGOs, all feel under attack today.

The speed of communications today means that more information, in both width and depth, is available to more people, faster, than ever before. People's faith in their governments has declined precipitously over the past fifty years, as it has in nearly all institutions. Even in Great Britain, mother of parliaments, confidence in British institutions has generally declined over the past three decades, and is now extremely low. Fewer than half the British public say they can trust their civil servants to tell the truth (47%), only a quarter of the public feel they can trust business leaders to tell the truth (28%), and fewer than a quarter say they have faith in the veracity of their government's ministers (21%). Fewer than half even say they trust what scientists working for either government (46%) or British industry (43%) have to say about the environment. Over the period there has also been a decline in confidence in industry, although the trade unions are no longer the bogeymen they were in their period of 'beer and sandwiches at No. 10', when two trade union leaders were thought to be among the most powerful half-dozen people in the country. Now they hardly rate among the so-called 'Power 300', among which Bill Gates is number three.

There has also been a precipitous decline in the confidence people have in the system of governance of the country, and there is a huge majority for such constitutional reforms as a Bill of Rights, a Freedom of Information

Act, and even a written constitution. The Jenkin Report observes that there is a crisis of 'trust', 'whatever precisely this may mean', it says. It is not difficult to criticize the Jenkin Report, but it would be foolish to dismiss it as meaningless. It should be read, considered, and acted upon by government, and extended holistically to include both the private sector and the civic sector, by academia and by the media, to a much greater degree than has heretofore been the case. As the Report concludes: 'Policy-makers will find it hard to win public support, or even acquiescence, on any issue with a science component unless the public's attitudes and values are recognized, respected and weighed in the balance along with the scientific and other factors'.

There is a four-stage process of effective communication, starting with awareness, the provision of knowledge, a feeling of openness, and the belief that the information is provided without any 'hidden agenda', and from a source of trust. The second stage is involvement, where the individual can see some clear link between themselves and/or their family, can in some way benefit, be made healthier, or richer, or better-feeling in some way. The third stage is persuasion, in that the individual feels informed and aware, and alert and involved, and is in a receptive mood to listen to the argument. The fourth stage, then, is action: to do what the giver of the information wishes to be done, whether to quit smoking, or diet, or exercise, or cut energy use, or use their car less, or whatever action the communicator wishes the recipient of that information to do.

If that is kept in mind, and scientists accept that people have the right to know for whom they are acting, the right to know what their scientific studies have concluded, and a feeling that they are being treated as responsible citizens, then bridges can be rebuilt. It is unlikely that the blind faith in the men in white coats will return, so expect that in the future scientists will have to take the time and trouble to explain what it is that they are trying to do, how they are going about it, and who will benefit therefrom.

Appendix

General public surveys

The details of the surveys of the general public cited in this paper are as follows. In each case, the survey data were weighted to match the known profile of the national population.

Multi-client co-operative survey on Business and the Environment: MORI interviewed a representative quota sample of 3900 adults aged 15+ across Great Britain. Interviews were conducted face-to-face, in-home, on 24–27 September and 8–12 October 1999, as apart of MORI's regular

Omnibus survey using CAPI (computer-assisted personal interviewing) technology.

In the British fieldwork for the 1999 International Environment Monitor, MORI interviewed a representative quota sample of 975 adults aged 15+ across Great Britain. Interviews were conducted face-to-face in-home, on 7–10 May 1999, as part of MORI's regular Omnibus survey using CAPI (computer-assisted personal interviewing) technology. Clients included Greenpeace International.

For the Better Regulation Unit of the Cabinet Office, MORI interviewed 1015 members of the People's Panel aged 16+ across Great Britain, face-to-face in-home on 9–19 January 1999.

For the Office of Science and Technology (OST) of the Department of Trade and Industry, a representative sample of 2200 members of the People's Panel was selected, of which MORI interviewed 1109 adults aged 16+ face-to-face, in-home, across Great Britain and Northern Ireland on 13 March–14 April 1999. The quantitative survey was accompanied by qualitative research, for which MORI conducted six two-day workshops around the United Kingdom between 5 December 1998 and 6 February 1999. In total 123 respondents attended the workshops. Three workshops were held in England, one in Scotland, one in Northern Ireland and one in Wales.

For the British Medical Association, MORI interviewed a representative quota sample of 2072 adults aged 15+ throughout Great Britain across 156 constituency-based sampling points. Interviews were carried out using CAPI (computer-assisted personal interviewing) face-to-face in respondents' homes on 3–7 February 2000.

For the Cancer Research Campaign, MORI interviewed a representative quota sample of 1933 adults aged 15+ across Great Britain. Interviews were conducted face-to-face, in-home, on 9–12 May 1997, as part of MORI's regular CAPI Omnibus survey.

For the Technical Change Centre, MORI interviewed a representative quota sample of 1824 adults age 15+ across Great Britain. Interviews were conducted face-to-face, in-home, on 4–9 June 1985 as part of MORI's regular Omnibus survey.

For *New Scientist*, MORI interviewed a representative quota sample of 2009 adults aged 15+ across Great Britain. Interviews were conducted face-to-face, in-home, on 5–8 March 1999, as part of MORI's regular CAPI Omnibus survey. The survey was published in the edition of 22 May 1999.

In MORI's 1999 Survey of Environmental Journalists, 29 journalists from the national and regional press, specialist press, and broadcasting organizations were approached, of whom 2 were unavailable during the interviewing period; 25 were interviewed (a response rate of 93%). Interviews were conducted face-to-face on 30 September–20 October 1999.

For the Pew Center, 1762 interviews were carried out in the USA and c. 950 in other countries (1997).

For the Medical Research Council, 1014 interviews were carried out among adults aged 15+, in Great Britain, on 1–26 September 1999.

Surveys among scientists

Random sample of 1540 scientists across 41 universities in Great Britain and among 112 scientists working for Research Councils were interviewed face-to-face among carried out between 13 December 1999 and 24 March 2000 for the Wellcome Trust.

Further details of these and other relevant surveys are to be found on the MORI web-site, www.mori.com.

Acknowledgements

Thanks to my colleagues, Dr Roger Mortimore, Michele Corrado, John Leaman, Kay Wright, Robert Cumming, Brian Gosschalk, Gideon Skinner, and others, who take an interest in our work on Science and Society, and to American colleagues Humphrey Taylor of Louis Harris & Associates, to Andy Kohut of the Pew Center, to Dr Richard Wirthlin and James Granger of Wirthlin Worldwide, and to Carlos Elordi and the Roper Center at the University of Connecticut at Storrs, for providing access to comparative data from the USA. Finally, not least, to our clients who enable us to conduct the study of Science and Society in the first instance, and give us permission to publish the results to the benefit of all, including the Cabinet Office, Office of Science and Technology, DTI, Medical Research Council, *New Scientists* magazine, *Ecologist* magazine, *The Times*, *The Sunday Times*, the British Medical Association, and others.

References

Bodmer, W., The public response to GMOs, *Oxford Magazine*, Eighth Week, Hilary Term, 2000.

Corrado, M., Begin by listening, *Science and Public Affairs*, August 2000, pp. 12–13.

Hargreaves, I. and Ferguson, G., Who's misunderstanding whom? Bridging the gulf of understanding between the public, the media and science, Economic and Social Research Council, September 2000.

House of Lords' Select Committee on Science and Technology, Science and Society, HL paper 38, 23 February 2000.

New Scientist, Let the people think, 22 May 1999, pp. 26–31, 60–1.

Worcester, R., Why do we do what we do? A review of what it is we think we do, reflections on why we do it, and whether or not it does any good, *International Journal of Public Opinion Research*, **9**(1), Spring 1997; 2–16.
Worcester, R., Seeking consensus on contentious scientific issues, *Technology Innovation and Society*, Winter 1999; 14–24.
Worcester, R., Softening up hard science, *Science and Public Affairs*, June 2000, p. 7.

ROBERT M. WORCESTER

Robert Worcester is Chairman of MORI (Market & Opinion Reserch International). His principal research interests are in British politics, in attitudes and behaviour regarding the environment, and in attitudes to science. His latest book, with Dr Roger Mortimore, is *Explaining Labour's Second Landslide* (Politico's), and previously, with Professor Sam Barnes, *Dynamics of Societal Learning about Global Environmental Change* (ISSC/UNESCO). He served as Senior Vice President of the International Social Science Council/UNESCO, and as a member of the initial Human Dimensions of Global Environmental Change (HDP) Committee. Professor Worcester is past president of the World Association for Public Opinion Research (WAPOR). He holds a visiting professorship of government at LSE and is a Governor. He is a member of the Science & Society Committee of the Royal Society, and holds three honorary doctorates.

The information age: public and personal

JOHN M. TAYLOR

Introduction

Information is the stuff of life, from the DNA code to the signals received by the Hubble space telescope from the beginning of time. Electronic information created by humans has only been around since the invention of the electromagnetic telegraph in 1837—one of the first applications of Michael Faraday's discovery of electromagnetic induction in 1831. We have had a century of electronics and telecommunications, and fifty years of the transistor and the computer. But in just the past five years we have seen an unprecedented implosion of digital computing and communications to produce the critical mass of the World Wide Web. We are at the threshold of the information age in which the global information infrastructure will transform most aspects of our lives from how we work and learn, how we receive health-care, how we spend our leisure, and how we keep in touch with the people in our social networks, to the products and services delivered to us by industry and commerce. And it has already transformed the way we do science—from the human genome project and the environmental sciences, to physics and astronomy.

My aim in this discourse is to show you what is on the agenda for the next steps in information technology, electronics, and communications—what I shall call *informatics*. I hope you will become familiar with some new words and ideas, and I hope to show you some things you haven't seen before and demonstrate some quite new capabilities. This talk is intended to be about technology, not futurology, and about key ideas and principles, not just wizzy demos and gadgets. To borrow Martin Rees's phrase about science in the twenty-first century, this talk is going to be about the large, the small, and the complex.

In the context of informatics it is going to be about the public, the personal, and the communal—about information utilities, information appliances, and extended communities. And from the perspective of all us users, citizens, and ordinary people, it will be about what goes on behind the wall, in front of the wall, and through the wall. In the tradition of these discourses and this place we are going to try and do demonstrations both in the real world and the virtual world of cyberspace—both are still experiments and both are equally likely to go wrong!

The information utility

Let's begin by talking about what's behind the wall—in the domain called *cyberspace* (Fig. 1). It comprises a hug global communications infrastructure called the *Internet*, itself consisting of millions of computers and switches, to which are connected tens of millions of host computers or servers, and hundreds of millions of client computers or PCs, through which users access the services and other users.

Let me just remind you how fast informatics has changed in the past five years. Moore's law says that the price performance of semiconductor products doubles every 18 months. This applies to microprocessors and, just as importantly, to electronic storage and communications. Physicists

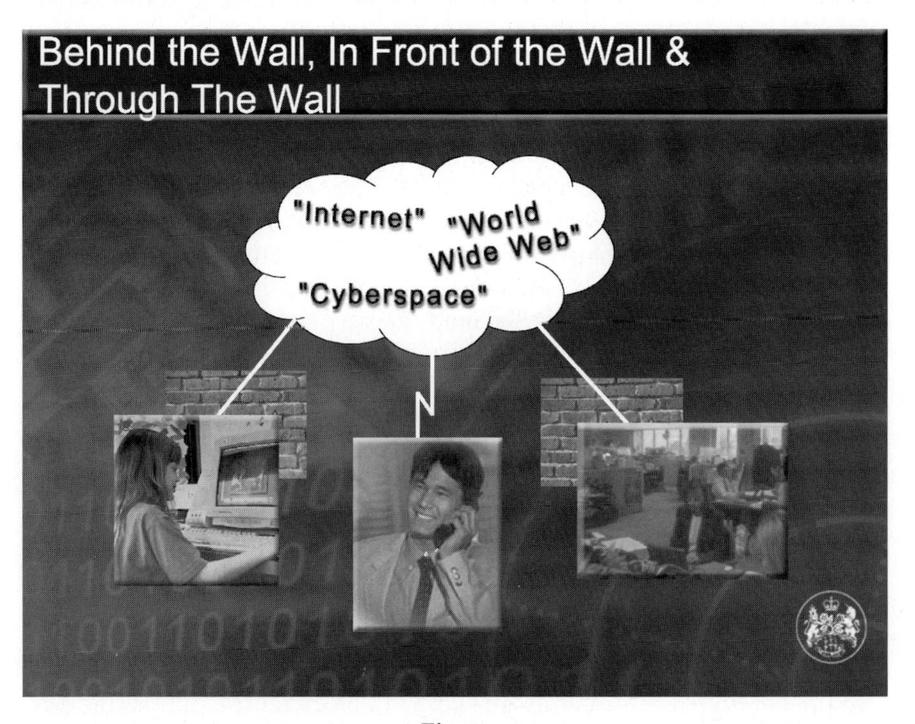

Fig. 1.

say it has another twenty years to run. Jim Grey's version of Moore's law in his 1998 Turing lecture[1] is that the progress in next 18 months will equal all previous progress. That is, the new storage added in the next 18 months will equal all existing storage on the planet, and the new processing power added in the next 18 months will equal all existing processing power on the planet. It is only approximate, but it makes you think.

The explosive rate of growth of the Internet and the World Wide Web is legendary. The number of host machines or servers on the Internet has increased exponentially over the past ten years. The number of hosts is around 74 million and the number of web-sites presently increasing at 60 000 per week (http://www.netsizer.com/). So, by the end of this year, the number of host computers on the Internet will have more than doubled over the past two years to around, say 100 million; the number of client PCs will have more than doubled to around 250 million. The amount of storage per host and per PC has gone up much faster, thanks to Moore's law. So the amount of storage, stored digital 'stuff', on the Internet, has probably gone up by more than 20 times over the past two years. The services available from all these servers behind the wall cover a very wide range from e-mail to huge image databases of space missions to pop music, as will see in a moment. Many of them are provided via the Interface of the World Wide Web which introduced the notion of the web-site as a place in a host computer to which a user could go, at which information is easily available, and through which many other web-sites can be visited.

Let us start with earth imaging. You have probably heard about the companies starting up which offer the ability to order up over the Web a customized image from space, and have it sent to you over the Web within a few hours. One of the companies involved in the technology for this is Surrey Satellite Technology (SSTL), a spin-off from Surrey University. They have specialized in faster, better, cheaper earth observation satellites. The images we will see are from a satellite they built for about £5 million, and they compare well with images from the US LANDSAT satellite which cost £100 million (http://www.sstl.co.uk/dgrc). We can see Beijing at a resolution of 30m, from a satellite at 640 kilometres. These are false-colour multi-spectral images. Red means healthy vegetation, dark means urban use. You can see the moat around the Forbidden City. We can zoom in to see detail at 10m resolution.

We don't have to confine ourselves to earth imaging, of course. Thanks to the NASA image web-sites we can look at the surface of Mars as seen from the cameras of the Mars landers as their mobile rovers moved around the surface of the planet.

Figure 2 shows images are from the Sojourner Rover that landed on 4 July 1997. The next step would be to combine these images into a

Mars Rover

Fig. 2.

virtual-reality system on a web-site that you can control over the Web so you can look wherever you want to.

What we have seen so far are all recent images, kept as data in web-sites maintained on servers. There has also been an explosion recently of so-called real media and streaming sites, which can just feed you audio and video streams in real time. So now you can start a radio station on the web without a licence and without any spectrum or expensive transmitters—there are about three thousand such radio stations on the web today and many TV stations too. You can even see the BBC TV news in real time whenever you want it (http://news.bbc.co.uk/olmedia/video/news169.ram).

Some of the servers on the Internet are themselves connected back to real sensors, instruments, and other things in the real world. This gives a new dimension to cyberspace: the ability to access remote resources in the real world and then to control them, from cameras on motorways, to astronomical telescopes, to robots under the sea, up a volcano, or in orbit. We cannot do telepresence in orbit yet but we can go and look through real Web cameras, webcams, and really use images as data. Imagine when there will be a billion cameras on the Internet. So you can go and look at the snow in Davos or the weather in Edinburgh. Or check if your car is still outside your house when you are away, or look in your elderly mother's kitchen to see why she's not answering the phone.

Today is the start of Science Week. Actually it is now Science 'ten days'—maybe another case of Internet inflation! One of the many participants in Science Week is the new @Bristol Media Centre in Bristol. This is a webcam that we can control live over the Web, looking at the new Wildscreen World building in Bristol as they are installing the exhibits and preparing for the formal opening in June. This will be one of the foremost media centres in the regions (http://www.at-bristol.org.uk/pages/01about/about_frame.htm). Today is 17 March—which is St Patrick's Day. There is a webcam looking at the St Patrick's Day parade in New York. There the time is about 3.30 in the afternoon and we can drop in on the festivities (http://www.mte.com/webcam).

More seriously, all an astronomical telescope does these days is to focus photons on to a digital imaging chip so there is no real reason to be at the telescope at all—we can just put it on the Web and use it as a webcam. We can now look through a real telescope, such as the William Herschel Telescope at Las Palmas in the Canary Islands, in real time, not just web pages of images prepared earlier. There is webcam live in the control room and we can hear what is going on and speak live to the scientists (http://www.ing.iac.es/). It is actually quite an important day for them today. Their new INGRID infra-red camera had first light last night.

> *Hello, this is John Taylor at the Royal Institution. Can you tell us what we are seeing please?*

> *Hello, John. This is Rene Rutten at Las Palmas. As you know, last night saw the first light for INGRID—the Isaac Newton Group Red Imaging Device mounted on the 4.2 metre William Herschel Telescope. Things went very smoothly and we started to obtain images within 45 minutes of sunset. INGRID is performing extremely well and we can now see some first-light images with minimal post-processing, including a 1 second exposure of Jupiter in Br Gamma.*

E-science

The major driver of the present generation of this cyberspace technology was the science and technology research community around the world. The Internet started life as the ARPAnet in the 1970s and spread rapidly to link the science and technology research community around the world. In those days I worked with Bob Kahn and Vint Cerf—the inventors of the Cerf–Kahn protocol that we now call TCP/IP, the protocol that runs the Internet. We worked on linking the UK research networks to the US ARPAnet in order to create the first stages of the international Internet. And everybody knows the World Wide Web was invented in 1990 by an Englishman, Tim Berners-Lee, working in the CERN centre for fundamental physics research in Geneva.

What has happened over the past few years as a result is the beginning of what I shall call *e-science*. That is, the use of the cyberspace infrastructure to enable large global teams of scientists to collaborate on fundamental research in ways that were just not feasible even ten years ago. It does have the disadvantage of producing papers like the OPAL paper from CERN, with 320 authors from 34 institutions.

So now I want to show you a few examples of e-science, global collaborations to do science. In particular, we will look at 'big physics'—the LEP and LHC at CERN—and 'big bioscience'—sequencing the human genome, and post-genomics and protein structures. I apologise in advance to the environmental science community and others that time does not allow me to include their equally compelling examples. These global scientific cooperations need to acquire huge amounts of data, analyse it, and combine it, and do huge computer modelling studies *in silico* to develop and test the theories to explain what they observe.

What will emerge as we go through this is that the present Internet infrastructure is already hugely inadequate—by a factor of at least a thousand times. This is driving the invention of the next generation of information infrastructure 'behind the wall'—the move from today's architecture of clients talking to individual servers and web-sites to what we should think of as an information utility, by analogy with the power, water, and even phone utilities. These provide pervasive services we take for granted. They have unquestioned levels of reliability, accessibility, and quality, without us ever having to worry about how they are organized behind the socket or the tap on the wall. But the new information utilities will be hugely more complex and interesting than the water utility. They will comprise resources of processing power, information storage, and communications, each thousands of times more powerful than what is available today. They will be combined in new ways to provide the resources and services users want without them having to know about where they are or where they come from. To paraphrase Jack Dongarra from Oak Ridge Labs,[2] the aim is provide users with the illusion that they have at their desk top a supercomputer ten times more powerful than the best in the world, and personal copies of the biggest databases on the planet.

Units

So before we move on, I need to remind you about some new units for storage, processing, and communications. For storage we move from gigabytes to petabytes and then exabytes. For processing power we go from MIPs to teraflops and then petaflops. For communications, from megabits to terabits per second (Fig. 3).

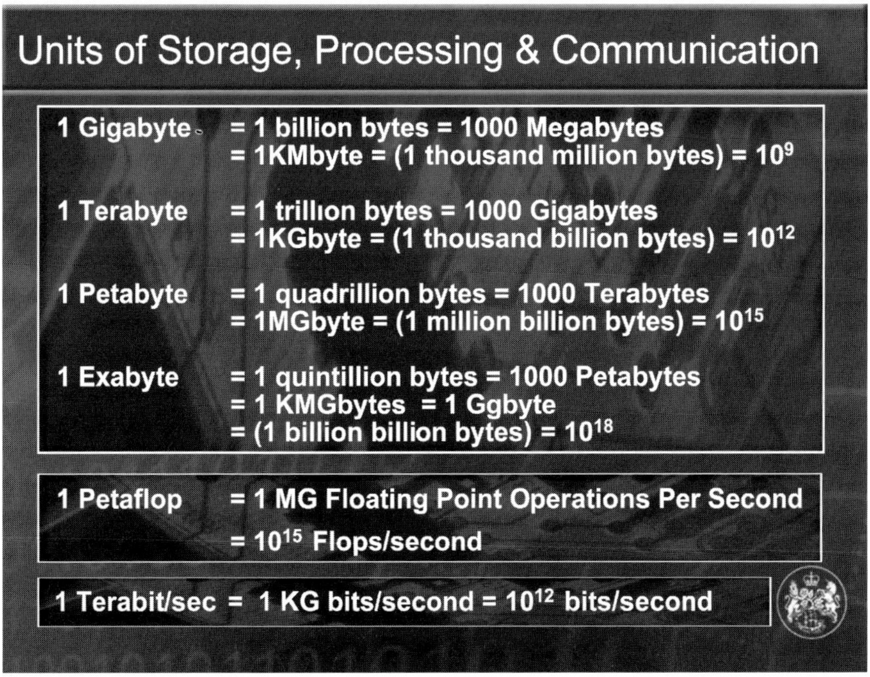

Units of Storage, Processing & Communication

1 Gigabyte = 1 billion bytes = 1000 Megabytes
= 1KMbyte = (1 thousand million bytes) = 10^9

1 Terabyte = 1 trillion bytes = 1000 Gigabytes
= 1KGbyte = (1 thousand billion bytes) = 10^{12}

1 Petabyte = 1 quadrillion bytes = 1000 Terabytes
= 1MGbyte = (1 million billion bytes) = 10^{15}

1 Exabyte = 1 quintillion bytes = 1000 Petabytes
= 1 KMGbytes = 1 Ggbyte
= (1 billion billion bytes) = 10^{18}

1 Petaflop = 1 MG Floating Point Operations Per Second
= 10^{15} Flops/second

1 Terabit/sec = 1 KG bits/second = 10^{12} bits/second

To get some feel for what these units mean, an ordinary personal photograph takes at least 1 megabyte to store digitally. A diskette holds about 1.5 megabytes—that is, about one photo. A CD-ROM holds 670 megabytes, less than 1 gigabyte, which is about a music album. A DVD holds around 4.5 gigabytes, which is enough for a full-length movie. Certain DVDs that are double-sided and double-layered hold about 18 gigabytes, which will give you sound in several languages, stories about the movies, and just about anything else you might want. There were about 20 billion photos taken in the United States in 1998. So world-wide there were at least 50 billion. At 1 megabyte per photo, this would take about 50 billion megabytes, which is 50 terabytes.

Michael Dertouzos at the Massachusetts Institute of Technology (MIT)[3] suggested another unit, the Library of Congress, or LOC (Fig. 4). The Library of Congress is the largest library in the world and he estimated that all the textual data it contains to be about 100 terabytes, which corresponds to about 200 million books. This is 1 LOC and it is about the same as the storage needed for all the 100 billion personal photographs probably taken world-wide last year. It would probably take about 10 000 LOCs to store everything ever written, printed, composed, and performed in all the cultures on earth—which is 1 exabyte, or 1000 petabytes. For comparison, all of the BBC TV film archive is about 2 petabytes.

That was about storage. If you want to know what a teraflop looks like, the most powerful publicly admitted computer in the world last year was the ASCI Red 1.8 teraflops machine (Fig. 5). It uses Intel chip technology and its main purpose is *in silico* nuclear weapons testing.

Global collaborations on science

So let me show you a little about some e-science projects.

'Big physics'

At CERN the LEP machine produces data at 2000 terabits per second from instruments like the Babar calorimeter (Fig. 6) on the Stanford Linac machine, completed last year. CERN uses these instruments to study collision events—what happens when you smash different kinds of particles together at huge energies. One of their most recent publications has been about smashing heavy ions together in another machine called the SPS, which involves capturing and analysing events in seven different detectors concurrently.

The new Large Hadron Collider machine is due to come on stream in 2005 to continue the search for the Higgs boson and an understanding of

Library of Congress

1 LOC= all textual data in US Library of Congress
= 200 million books = 100 Terabytes = 0.1 Petabyte

10,000 LOCs needed to store everything ever written, composed, performed, painted, filmed or recorded so far
= 10K .100 Terabytes = 1M Terabytes = 1K Petabytes
= 1 Exabyte

All BBC Film & TV Archives = 2 Petabytes

Library of Congress

The Library
of CONGRESS

10,000 LOC = 1 exabyte = everything ever written, composed, performed, painted, filmed or recorded so far

If number of servers on Internet today = ?100 million
and hard disc per server = 400 G bytes
If number of PC's on internet today = ? 500 million
and hard disc per PC = 20 G bytes
Then online storage on internet today = 50KMG bytes
= 50 K Petabytes = 50 Exabytes = 500 K LOC

Fig. 4.

Fig. 5.

mass. Overall, the LHC will produce raw data at 1 petabyte per second, which is 1 million gigabytes per second, or 1 million CD-ROMs per second. This is instantly reduced by dedicated silicon and software filters by a factor of 10 million to 100 megabytes per second. These experiments are about searching for an incredibly tiny needle in a huge haystack. They will each need to run for up to 24 hours and the filtered data for a 24-hour run is 10 000 gigabytes or 10 000 CD-ROMs. So one year's worth of filtered data is about 1 petabyte, 1 million gigabytes, which is 1 million CD-ROMs—which at five to the inch would be about three miles high! This has to be kept as the prime output of these very expensive experiments, carefully looked after in big databases and made accessible to scientists all around the world as the prime source of data for their research. This is why we are starting to talk about data curation, not just database management.

So the computing, storage, and communications infrastructure to support the LHC is as important as the LHC machine itself. It is no good having one without the other. In fact, you could really say that the main function of the LHC itself is to shovel bits into a server for scientists all round the world to use as the prime source of data for their research. And today's Internet, Web, and client–server architecture is quite inadequate. It does not pass Jim Grey's scalability test—architectures that will scale up by a factor of a million—in processing power, storage, and communications. So we need to have new kinds of

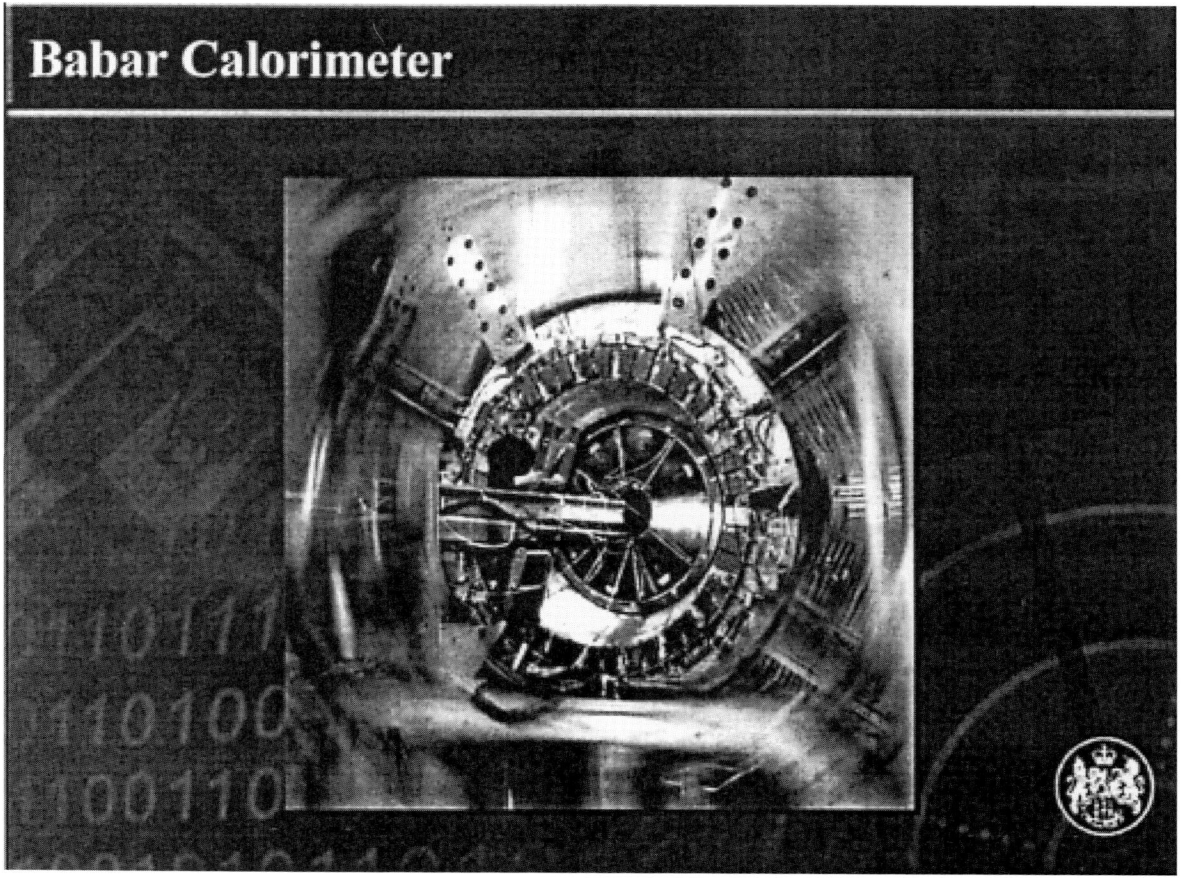

Babar Calorimeter

Fig. 6.

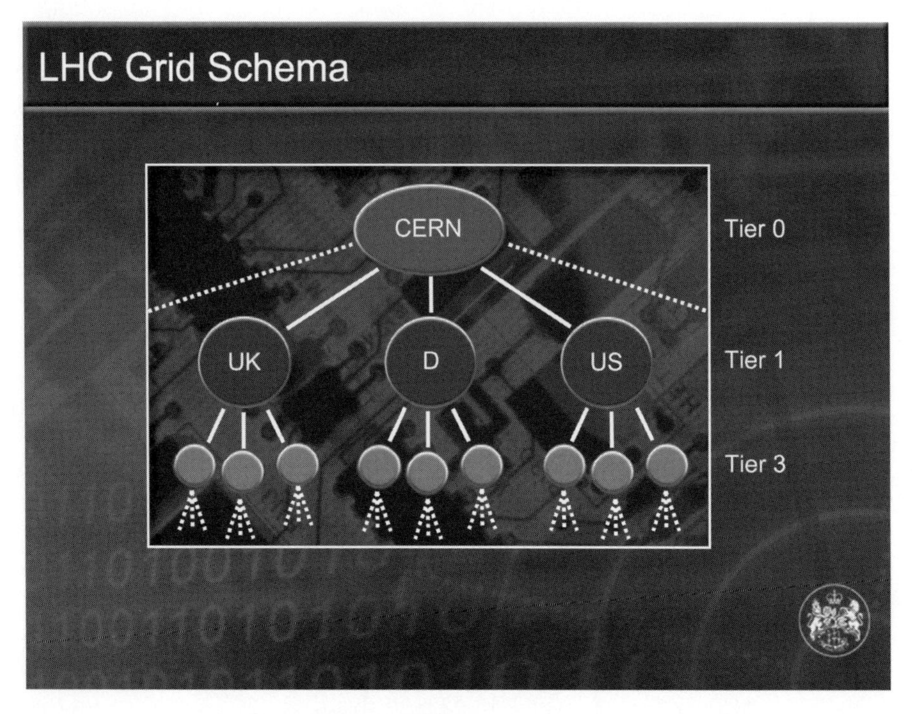

Fig. 7.

systems architecture 'behind the wall'—the first steps towards an information utility.

An important move in this direction is known as the GRID architecture, by analogy with electric power grids. It is being developed around the ideas in the so-called GRID bible from Ian Foster from Argonne National Lab and Carl Kesselman from USC-ISI.[4] A particular version of a GRID architecture is being developed for the LHC (Fig. 7). This involves four tiers of systems—Tier 0 at CERN itself in Geneva, at Tier 1 node in each of the major countries, and tier 2 nodes within the major countries, and so on.

The basic scheme in the architecture has four levels (Fig. 8): the underlying communication resources, the actual computation and storage resources, above that the information level, and above that the knowledge level. All this supports successively more powerful abstractions of the basic data and computation structures. Much more computational science research is needed to understand how these should be organized.

Just to give you a quick glimpse of some of the elegant technology at the very foundations of these systems (where we in the UK are outstanding, as it happens), there exists a little miracle called the EDFA—the erbium-doped fibre amplifier. It illustrates a key leap of technology

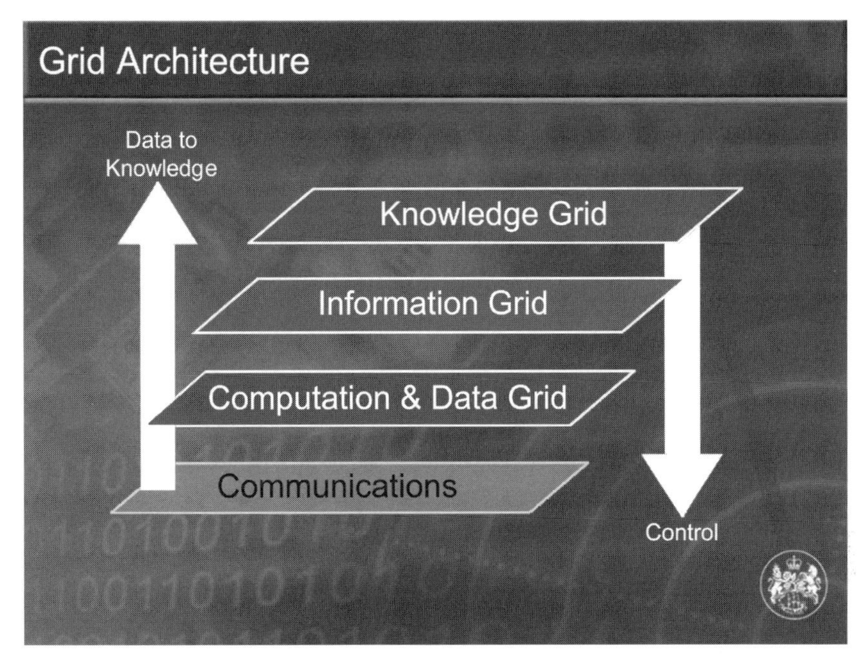

Fig. 8.

which the communications layer has to make in the next generation, to get the electronics out of opto-electronics and go to all-optical transmission and switching. The EDFA is an all-optical amplifier for fibre optics invented at Southampton University. It has dramatically simplified and cheapened the fibre infrastructure.

Bioscience

Just in case you think the physical sciences are having all the e-science fun, let us now turn to biosciences, and as a particularly topical example of e-science, the Human Genome Project.

The Human Genome Project is an international 13-year effort to decode the human genome. It started in 1990 and involves scientists in the US, the UK, Japan, France, Germany, and China in at least 25 major institutions. There are 23 pairs of chromosomes responsible for carrying the human genome (Fig. 9). They are estimated to contain about 100 000 genes, which are encoded in about 3 billion base-pairs. The sequence data as it is obtained in all the collaborating laboratories is pooled in the genome database, or GDB, the public global data repository for human genome mapping. According to the GDB, about 66.8% of the 3 billion base-pairs have been done so far. The project is aiming to have a rough

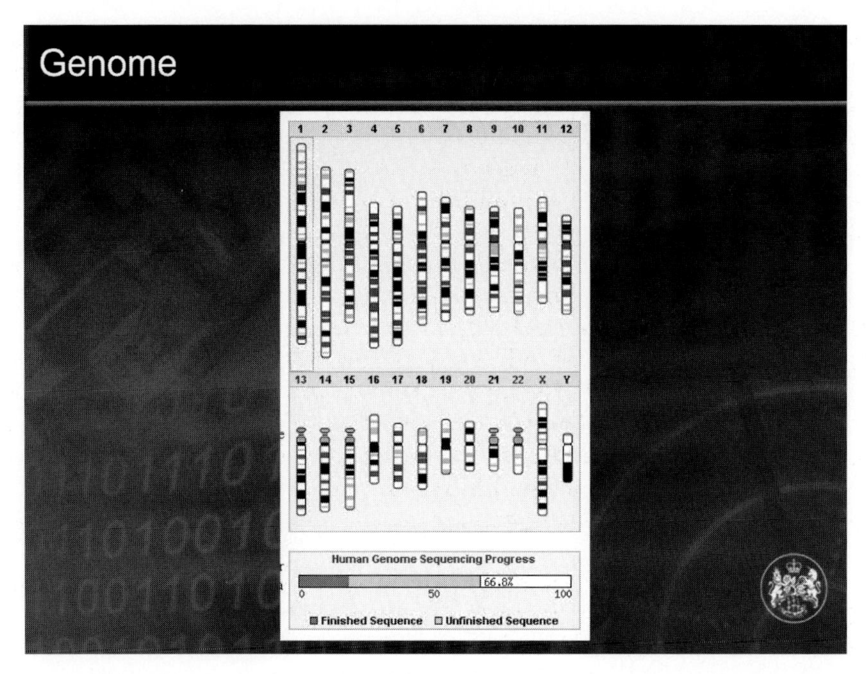

Fig. 9.

draft of the sequence 90% complete by the summer of this year (1999), which is about two years earlier than people expected two years ago. They aim to have a high-quality version finished by 2003.

The leading laboratory in the UK is the Wellcome Trust Sanger Centre in Cambridge. On 1 December 1998 the complete sequencing of the first human chromosome, Chromosome 22, was announced by a joint team from the Sanger Centre, the Universities of Oklahoma and Washington in the USA, and Keio University in Japan. Gene sequencing involves hundreds of machines working 24 hours a day, 7 days a week, again putting all their results into servers on the Internet. Indeed, you could again be unkind and say the main function of labs like this is to put bits into servers on the Web for use as primary data by scientists everywhere. Sequencing the genome of one person is only the start. Each of the 100 000 genes codes for a protein and the structure and function of many, if not most, of these proteins is unknown.

One particular area of study is so-called protein folding. Proteins are complex three-dimensional structures containing ribbons, helices, and many other features. The way they curl up in three dimensions creates intricately shaped nooks and crannies, which are crucial in understanding and predicting their behaviour. Fifty-per cent of human cancers involve mutations to protein p53, a DNA-binding protein involved in controlling cell division, which prevent it binding to the DNA. Some of the most demanding computational science going on anywhere at present is trying to model these folding processes *in silico* to produce videos like those developed by Julia Goodfellow at Birkbeck College.[5] In real life, the folding takes about 20 microseconds and this simulation represents only about 1 microsecond. Nevertheless, to generate a simulation took 2 months of computation on the Cray T3 supercomputer at Manchester.

Computational bioscience is why IBM recently announced its Blue Gene program[6] to get to petaflop-scale computing in about five years—a thousand times more powerful than the ASCI Red machine I showed earlier (Fig. 10). Blue Gene will consist of more than 1 million processors each capable of doing 1 gigaflop. Thirty-two of these processors will be put on a single chip and 64 of these will then be mounted on 2 foot by 2 foot board to give 2 teraflops—the same as today's ASCI computers, which occupy 800 square feet. Blue Gene will be formed by putting eight of these boards in a rack and combining 64 of these racks to get the final 1 petaflop.

The one thing we can be quite sure of is that there will be plenty of bioscience for such machines to work on. Not only do we not know much of the structure and functions of the proteins coded for by most of the genes, we know little about how genes vary across populations and

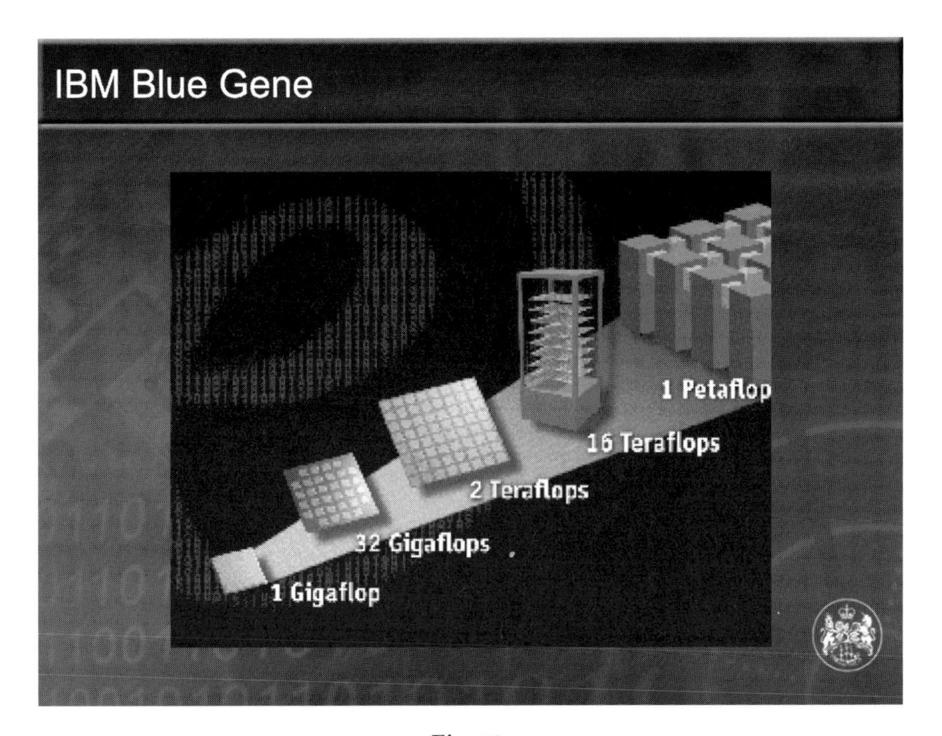

Fig. 10.

with diseases, or how genetic variations amongst people with a particular disease correlate with their response to different drugs and therapies. So there remains a huge amount of research and development to do before we can deliver mass customized prevention and therapy to a population the size of the UK's tailored to each individual's genome.

So the new information utility architectures will provide scalable systems that will pass Grey's 'times a million' scale-up test in both storage and processing power. They must allow thousands of scientists to contribute to giant global databases and tens of thousands of scientists distributed around the world to access and use the data in real time. This means going way beyond today's data mining and search engines. We need methods for data curation, quality assurance, provenance, and security, as well as software and middleware behind the wall to create and maintain the illusion of huge dedicated resources easily and cheaply available at the desk top. And, as we have said, DARPA built the Internet, CERN invested the World Wide Web, and this led to today's e-business revolution. The information utility infrastructure being designed now for e-science means it is going to happen again. This will generate the technology that will power the next generation of e-business and e-society.

Small, personal, 'in front of the wall'

Now let's move in front of the wall to the world of the small and the personal, to the world of personal information appliances—digital medial 'things'. I want to focus now, on the huge increases in the processing power and data storage, digital stored 'stuff', that I personally can own, embedded in all kinds of devices and places in my personal environment. I want to illustrate a few key ideas.

First, the idea of an information appliance. The devices available today for ordinary people to participate in the information society are confined to the phone, PC, TV, fax machine, and printers. As we enter the information age, it seems to me unlikely that this will be the last word. I am glad to tell you that there *is* life after the PC and Windows. We are seeing the cost of computer processing and storage moving right down to the consumer price point. It is now practicable to embed huge quantities of processing power and storage in small, useful, single-purpose devices that you would tolerate in your living room and personal space, more like information tools than general-purpose PCs.

A very important driver of this is personal digital imaging. We have had the phone, and then e-mail as forms of personal electronic communications that have come to be quite ubiquitous. Now we will be adding personal imaging—the ability to show someone an image or send someone an image and talk about it. Personal digital imaging is about to become as pervasive as speaking on the phone and e-mailing.

Another key idea about information appliances is that they are not computer peripherals like printers and scanners. They are stand-alone appliances that do not have to be plugged in to a PC before they can do anything useful. For example, the Hewlett Packard (HP) digital camera is a true digital appliance because it does not need to be plugged into a PC before I can do anything with the images it has captured. Instead, it is part of a family of digital media appliances that can interwork directly with each other when they need to (Fig. 11).

So, if we want some prints of the photos we have taken, we need a printer that is also a stand-alone information appliance. So all you have to do is point, shoot, and print. This works because any appliance in the family has point-to-point communications capabilities built in, in a rather subtle way. I know it's subtle because my labs in HP in Bristol spent several years developing the ideas, the technology, and the industry standards, so that an appliance made by one manufacturer can talk to one made by someone else, when it happens to need to. In the case of the digital camera and printer the appliances have infra-red communications at 4 megabytes per second and support the HP Jetsend appliance protocols.

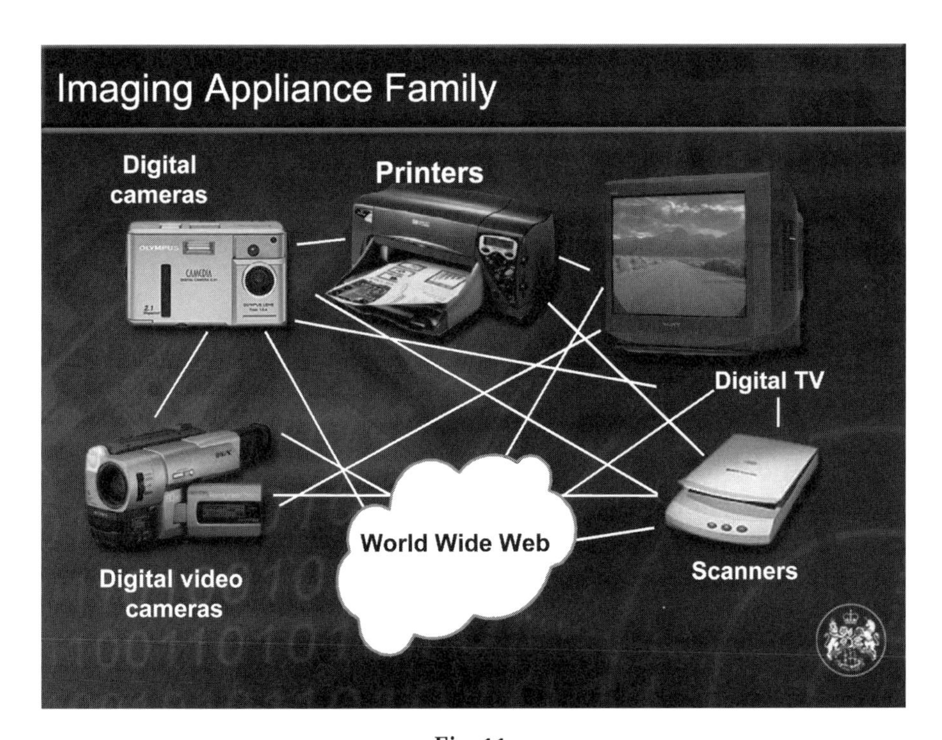

Fig. 11.

These demonstrations also illustrate the other two key points in the information appliance story. First, families of cooperating appliances represent a new kind of system. They form '*ad hoc* systems' for casual unconscious use by interacting with each other when they happen to need to. This is a very different kind of system from any we have been used to. The various devices in a family need to agree on how to interact with each other yet they will do so infrequently and unpredictably.

The other idea is simpler. Wireless communications is another revolution that is crucial both for short-range point-to-point appliance communications, as we have just seen around the home, and for the workplace. These used infra-red, but point-to-point radio is the other key modality. Look out for SWAP and Bluetooth for short-range low-power radio links around the house and the desk top and for body-nets for wearable computers.

And, of course, for wide-area long-range communications, we have had the explosion of mobile phone services: first analogue, then digital, and we saw in the start of the auction of UK licences for the third generation of digital mobile services, UMTS. This will allow mobile phones to become video appliances and Web browsers, beginning with the wireless applications protocol, WAP, and moving to more powerful protocols later. The early WAP phones from Motorola can browse the Web as well as do all the normal things, including e-mail.

There are other information appliances in completely new categories of device. One example of this is called the HP Capshare, because it is for capturing and sharing images (Fig. 12). Its task is to let you do casual capture of anything on paper that takes your fancy, and then share it with other appliances in the family. So, if I wanted to make a copy of say a flipchart, all I do is to swipe the Capshare like a paintbrush over it and then point, shoot, and print it on the stand-alone Jetsend appliance laser printer.

By the way, this is a really smart paintbrush. It is able to stitch the image back together because it navigates by sensing the random fine texture of the surface of the paper and then correlating across the data set it captures. That requires a processing power of about 1 billion operations per second—a gigaflop—which is built in to this $300 product. If I want the image in my laptop which also has Jetsend infra-red interface, I can point, shoot, and save it there, and then view and e-mail it to anyone—even put it on my web-site if I wish.

Yet another new category of appliance is the experimental audio–photo player. The notion here is that the sound that was going on at the time you took a still photo conveys much of the atmosphere of the scene and the memory of the event. To test this, we tried various versions of how it might work. This one has the sound in a little chip bonded to the photo. So now I have my farewell presentation from HP Labs recorded in both image and sound format.

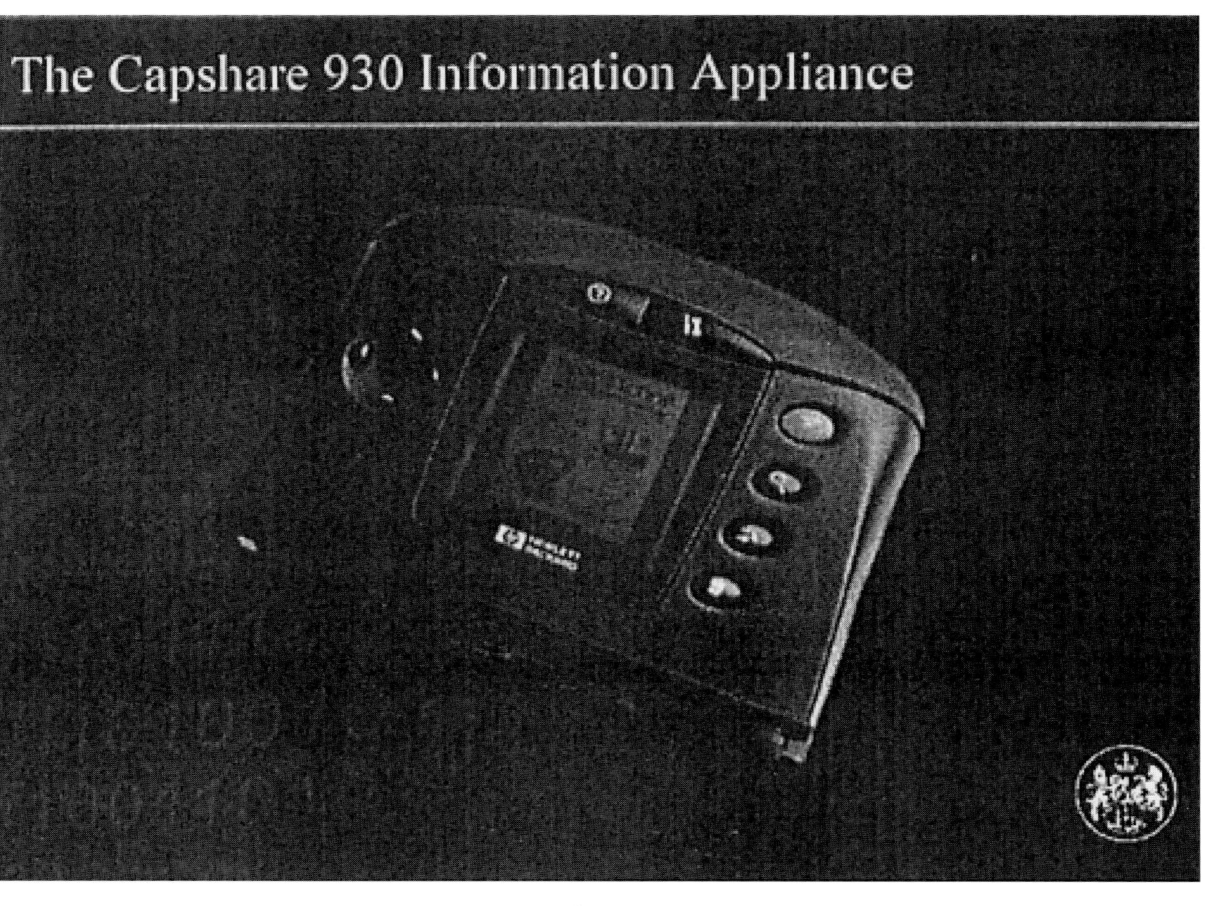

Fig. 12.

Of course, greeting cards using chips are already on the market. The worrying thought is that the storage and processing power in these cards is probably greater than that in the first valve computers of the 1940s.

Complexity, community, 'through the wall'

So, we have talked about the large and the small, the public and the personal, and 'behind the wall' and 'in front of the wall'. Let me finally turn to the third area—the complex, the communal, and what goes on through the wall. Again I want to illustrate a few key ideas.

I have already shown you one new kind of system—the '*ad hoc* system'—that forms transiently when appliances interact with each other. Now let me show you another new kind of system—'the *end-user system*'. This is the system formed when appliances in front of the wall work with services and resources in the utility behind the wall to produce a service or a capability that is greater than the sum of its parts.

We considered earlier the WAP phone. This is a wireless appliance that allows Web browsing; that is, it allows the user to access a whole range of services and facilities behind the wall. Future versions of such appliances will incorporate GPS chips, receivers, and processors, for the signals from the Global Positioning System satellites, which would let the WAP phone know where it is, anywhere on the planet, to within a few metres. It can then start to deliver what we call location-based services—getting information from the Web that is relevant to where the user happens to be. To explore this, we worked with Bristol University's wearable computing project and came up with, not the Laserjet, but the BlazerJet (http://wearables.cs.bris.ac.uk/).

BlazerJet can sense if there is anything of interest to the wearer nearby (Fig. 13). The computer in the lining of the jacket is connected to the Internet using a GSM mobile phone, and it has the GPS navigational system. It has built-in speech recognition, and uses a HP Jornada palmtop computer to display its output. Small short-range transmitters called 'pingers' have been developed; these can be placed at useful locations, in shops, bus stops, cafés, or on people. These pingers tell BlazerJet the addresses of web-sites associated with these places which may have interesting information for the person wearing the jacket.

So we made a short video to show what a BlazerJet might look like out and about and in action. In the video, Cliff Randall comes from Bristol to visit London and is looking for information about buses. His jacket senses the bus-stop's pinger, and tells him when the next bus to Piccadilly is coming. As he is walking along the street, the jacket senses a bookshop, sends his current shopping list into the shop, and tells him

Fig. 13.

that the shop has one of the books he is looking for. As he is walking down Piccadilly, the jacket matches the GPS location with the Royal Academy web-site. There is something of interest there, so he goes into the Arts Café. He orders a coffee and doughnuts using the Internet—no queuing needed—while he's in the café the jacket lets him know that there's an interesting lecture taking place around the corner in the Royal Institution. So in he goes, and, despite the locked-door tradition, here he is! And instead of shaking hands, our wearable computers can exchange our business cards automatically.

The latest end-user system to hit the headlines is MP3, of course. Apparently, MP3 has just superseded sex as the most searched-for word on the Web! For those of you that still don't know what MP3 is, it is the ability to down-load music tracks from the Web into a personal appliance called an 'MP3 player' (Fig. 14).

By the way, all this is going to give a further major push to the expansion of personal digital storage—my personal digital 'stuff'. If I send an e-mail, it is 1 kilobyte. If I attach a photograph, it is 1 megabyte. If I attach an MP3 sound track, it is 4 megabytes. A thousand-fold increase in the demand for personal communication and storage resources.

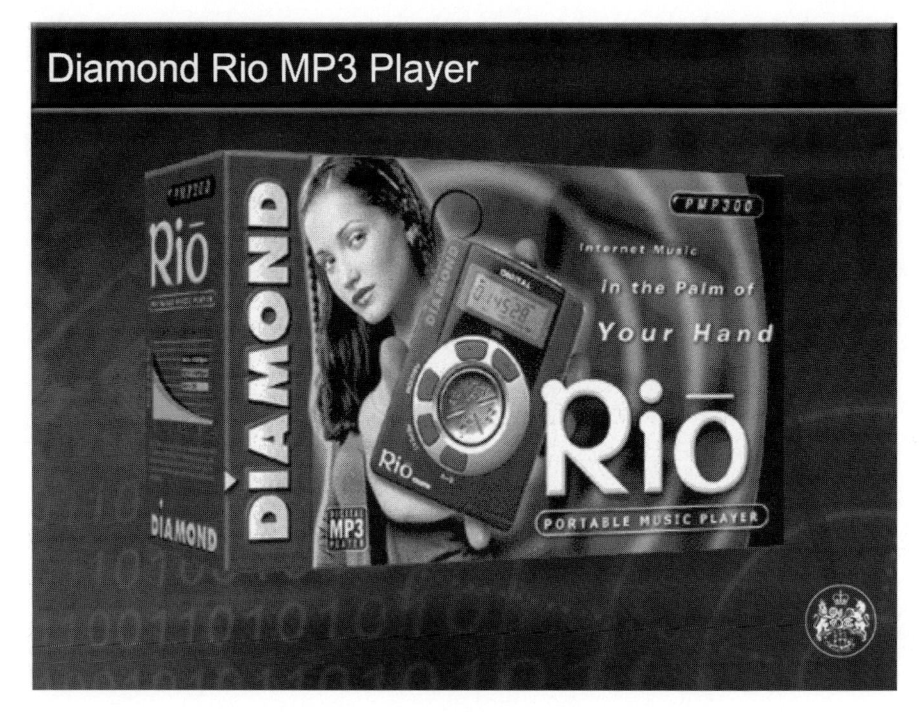

Fig. 14.

Agents and robots

The next step, of course, is to build '*ad hoc* systems' and 'end-user systems' which are intelligent and autonomous. We already experiment with primitive ideas of virtual agents in cyberspace, moving autonomously around the Web, searching and negotiating on our behalf to get the services and resources we want. In the 1970s there was an early flirtation with artificial intelligence, but people really had not done their sums.

Ray Kurzweil has estimated how much processing power it will take to get into the same ball-park as a mouse brain—about 100 gigaflops (Fig. 15).[7] And a human brain would take 100 000 times more, or 10 petaflops. That would be about 10 Blue Genes. So when we can do 1 megaflop for 1p, a mouse brain might cost £1000. According to Fig. 15, this might be in about 2020, which is getting distinctly interesting from an appliance point of view. At this level we can contemplate not just virtual agents in cyberspace but real autonomous agents in the real world with vision, speech, and the other key sensors and affectors they would need to survive. In short, we could have robots that would cost about the same as a PC.

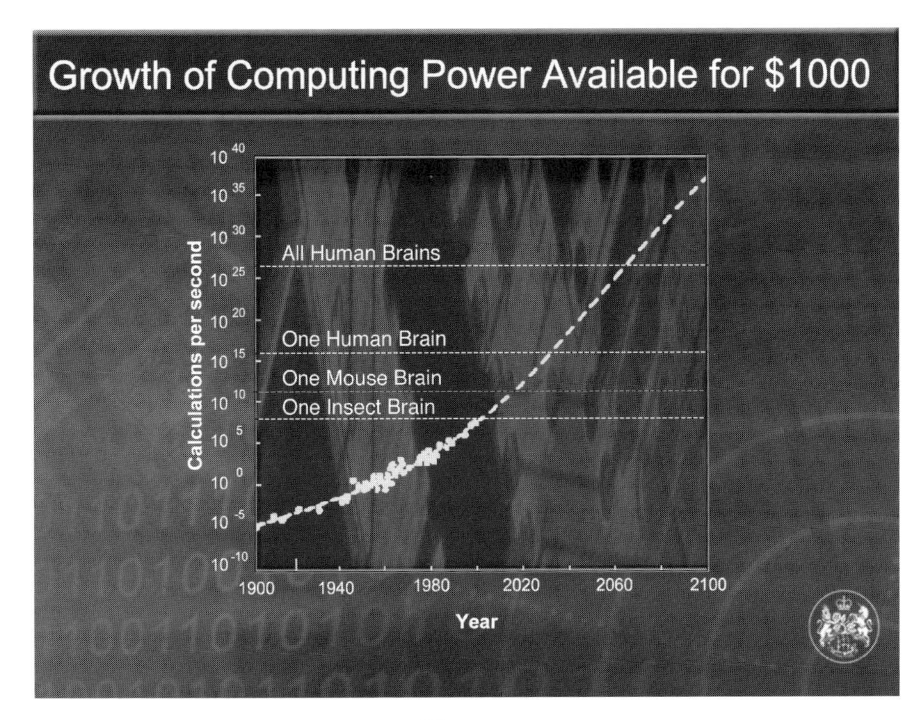

Fig. 15.

I wish I could say I had a prototype here but I think we have a little way to go yet. However, you might have heard of the Sony Aibo dog (Fig. 16). There are not many around at the moment. Apparently they have been having some problems about passports at immigration since the quarantine regulations were changed! This little dog cannot sustain your attention for very long, but it illustrates quite nicely that they could become available as domestic pets, servants, and companions. These might be useful, friendly, and fun, especially perhaps for the rapidly increasing number of very elderly, lonely, and handicapped people that we shall surely see as a result of developments in medical science over the coming decades.

And then, of course, there is quantum computing, DNA computing, and so on.

Community

And this takes me, finally, to the communal: to what happens when you reach through the wall and the information utility helps people to be in touch with each other, and helps the growth of quite new kinds of communities of people in the real world. It is already happening. Anyone

Fig. 16.

can have a web-site, and web-sites can be communal places where like-minded people find each other and get together. As well as all kinds of e-commerce and e-retail, we now have e-auctions that have enabled the development of many new global niche communities of interest to people involved with, for example, out-of-print books.

We have DIY e-health communities that are growing very fast where people are able to consult each other about their symptoms and the efficacy of their treatments in parallel with their involvement with their GP and the formal medical system. Its interesting that GPs are now offered training on how to deal with patients who have consulted the Web before coming to consult them. DIY over-the-counter health maintenance products and services will be a fast-growing complementary area to medicine as people spend more of their disposable time and income on staying away from the medical system for as long as possible. There are many, many e-education community projects, including, of course, e-university projects in the USA and the UK. Another aspect of the disruptions coming from informatics is e-democracy. Last week the US presidential primary in Arizona became the first public election to be conducted via the Web (http://www.election.com/political/arizona/ photos.htm).

We need to think carefully about how this radical extension of ordinary people's ability to form new communities of interest needs to be balanced against the opportunities this offers activists and special-interest lobby groups to greatly amplify their ability to seize attention and agendas and reduce the democratic voice of ordinary citizens.

However the most important community for most of us is still our family, and particularly these days, our extended family. The Taylor family's experimental e-family web-site will make it easier for us all to keep in touch.

But perhaps my favourite image illustrating the value of personal imaging in the extended family comes from some trials we did several years ago in HP Labs with extended families in the US. We gave them early prototypes of easy-to-use information appliances focused around personal digital imaging. One of these was a flat-bed scanner and one of our parents stood their toddler up on the scanner and added a caption: 'Guess who's walking?' and sent it to grandma 2000 miles away (Fig. 17).

Conclusions

Let me leave you with three thoughts.

- First, informatics is for fun as well as for serious things like e-science and e-commerce. It is for spending your personal disposable discretionary time, as well as being efficient at work.
- Secondly, *e* is for 'electronic', but from now on *e* is really for 'extended': extended company, extended science, extended university, extended democracy, and, most of all, extended community and extended family.
- Thirdly, do not forget that all of these information appliances still come with on–off switches!

Thank you for your attention.

Acknowledgements

I would like to thank those who helped to put this discourse together: Constantine Biller, Office of Science and Technology, London; Professor Michael Wilson, Council for the Central Laboratory of the Research Councils, Didcot; Rosalind Darby, Engineering and Physical Science Research Council, Swindon; Cliff Randall, Department of Computer Science, University of Bristol; Professor Tony Hey, Dr Peter Smith, and Dr Kirk Martinez, Department of Electronics and Computer Science, University of Southampton; Professor Julia Goodfellow, Birkbeck College, University of London; Bob Vincent, @-bristol, Bristol; Dr Paul Murdin, the British National Space Centre, London; Dr Rene Rutten, the Issac Newton Group of Telescopes, Las Palmas; Jeffrey Ward, Jeremy

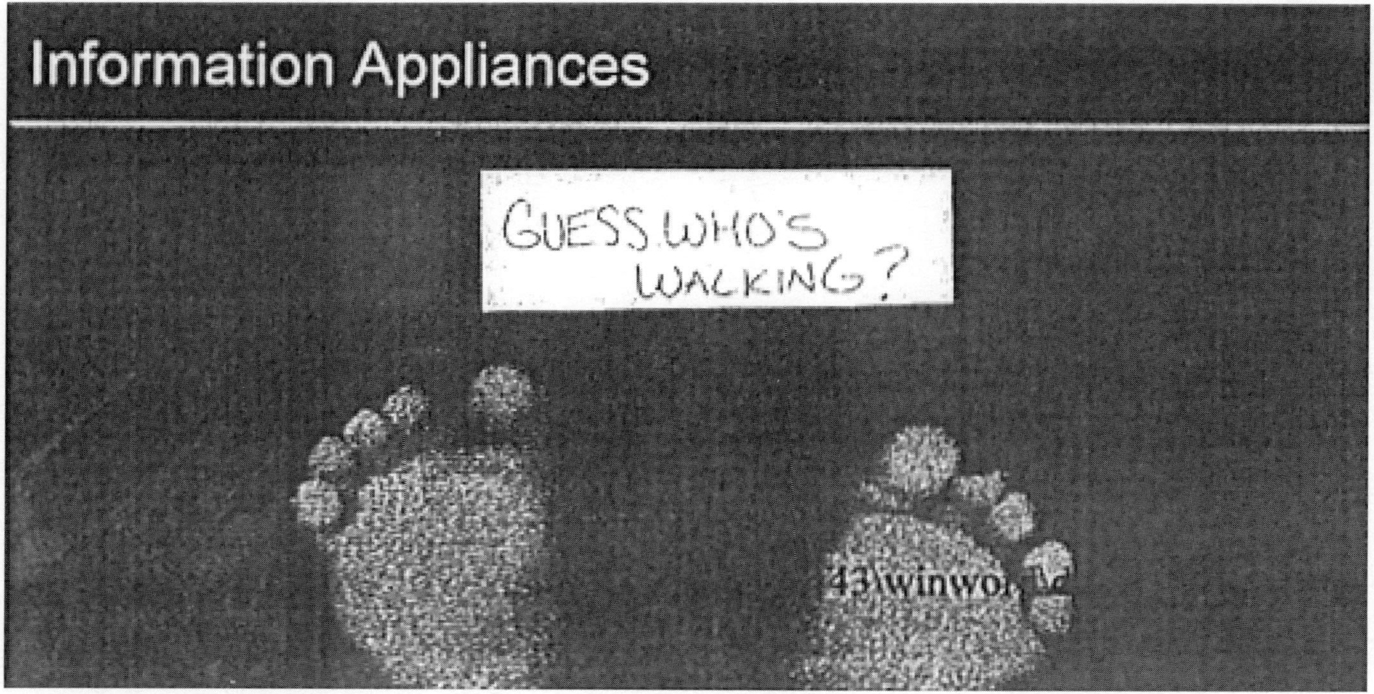

Information Appliances

GUESS WHO'S WALKING?

Fig. 17.

Crisp, and Audrey Nice, Surrey Satellite Technologies Ltd, Guildford; Andrew Wilson, the Institute of Electrical Engineers, London; Jo Reid, Hewlett Packard Laboratories, Bristol; the Particle Physics and Astronomy Research Council, Swindon; the Natural Environment Research Council, Swindon; the Medical Research Council, London; Bipin Parmar, Irena McCabe, and colleagues, the Royal Institution, London; Jonathan Stethridge and colleagues at Unique Media, Slough.

References

1. Jim Gray, 1998 Turing Lecture (Microsoft Research Advanced Technology Division: http://research.microsoft.com/~Gray).
2. Jack Dongarra, The marketplace for high-performance computers (Oak Ridge National Laboratory: http://www.ornl.gov/).
3. Michael Dertouzos (Massachusetts Institute of Technology Laboratory of Computer Science: http://www.lcs.mit.edu/).
4. Ian Foster and Carl Kesselman, The grid (Argonne National Laboratory: http://www.anl.gov/ and University of Southern California Information Sciences Institute: http://www.isi.edu/).
5. Julia Goodfellow (Department of Crystallography, Birkbeck College, University of London: http://www.cryst.bbk.ac.uk/).
6. IBM Research, Deep Computing Institute and Computational Biology Group: http://www.no.ibm.com/.
7. Ray Kurzweil, The age of spiritual machines: when computers exceed human intelligence (Kurzweil Technologies Inc.: http://www.kurzweiltech.com/).

JOHN M. TAYLOR

John Taylor is Director General of Research Councils in the UK Office of Science and Technology, responsible for the seven Research Councils funding research across the whole spectrum of science and technology in the UK science and engineering base. He was formerly director of Hewlett Packard's Laboratories Bristol the European arm of Hewlett Packard's worldwide long-range research laboratories, where he developed major programs of research in the areas of information utilities, information appliances and mathematics, including internet security, wireless communications, telecoms and personal digital imaging. Earlier, he led various research groups at RSRE and ASWE in areas including secure computing, communications, command and control. He was President to The Institution of Electrical Engineers 1998–9, and Chaired the UK Technology Foresight Panel in IT Electronics and Communications until December 1998. He is a visiting Industrial Professor at Bristol University and a visiting Professor at Imperial College, London.

Presenting science to young children

CHARLES TAYLOR

Introduction

Until about 25 years ago relatively few research scientists devoted any of their time to giving popular lectures to schoolchildren. Indeed, not only were very few involved, but those who were tended to be frowned upon by the scientific establishment for wasting good research time! The RI was, of course, a glorious exception, and successive directors ever since Michael Faraday have encouraged scientists to present lectures to children, and generations of children have come along to be enthralled. However, since the publication of the Royal Society's Report *The Public Understanding of Science* in 1985, more and more lecturers have been involved, but most of the lectures provided were for children of secondary school age. In the last fifteen years the RI lecture programme has included lectures to younger age groups and nowadays lectures to children as young as 7 or 8 are included. My intention in this discourse is to consider these changes, their justification, the modifications in lecturing technique they demand, and the reactions of children to the presentations.

My own involvement

Almost exactly 50 years ago I was a new assistant lecturer in the physics department at what is now called UMIST, but at that time we did not offer degrees in physics. All our courses were of the type 'physics for chemists', 'physics for engineers', etc. When it was decided that we should offer degrees in physics alone, we needed a public relations exercise to let schools know that it was available. It was decided to send lecturers out to schools to talk about physics, and I was one of the first to be 'volunteered' to go. In considering how to set about this task I remembered what an impression demonstrations had made on me during my own degree course. I was privileged to attend magnificent demonstration lectures by Sir Lawrence Bragg and by Dr Alexander Wood, among others. I particularly remember one by Dr Wood which involved Miller's Phonodeik apparatus which converted incoming sounds into oscillations of a small mirror. A spot of light fell first on this mirror and then on an octagonal arrangement of mirrors which stood on the turntable of an old 78 r.p.m. wind-up gramophone. As the mirror rotated, a graph of the waveforms was traced all round the walls of the old Cavendish lecture theatre in Free School Lane.

The demonstration, which was in 1940, made such an impression that I can still remember precisely the piece of music that he used—Rachmaninov's G-minor piano prelude. The spot of light bouncing around the walls in time with the piano made a lasting impression and I decided that physics and music might be a good subject for me to discuss and that I should attempt to use demonstration in the hope that they would make something of the same impression. So I set out with a van-load of demonstration equipment and visited many schools within a fifty-mile radius of Manchester. So began my interest in lecturing to children—but most of the lectures were to fifth and sixth forms.

Then, in the late 1960s, I moved to Cardiff and the reasons for lecturing to schools began to change. By that time there was a decline in university intake to science and it was thought that popular-level lectures might encourage children to consider a science career. But I began to think that, if the aim was to encourage children to take up science, it was a little late to leave it until secondary level, when most of the syllabus choices had already been made.

Change to younger age groups

The turning point came for me when I was scheduled to lecture to a group of children in the lower half of a secondary-level school in South Wales and as I was loading my equipment to set out I received a call from the school asking if they could bring in children from the junior school which was on the same campus. I agreed to the inclusion of some children aged 10 and 11. The teachers pressed me to include even younger children and I finally agreed to come early and to put on a special lecture for about a hundred 7-, 8- and 9-year-olds.

I was introduced to the children and began my talk. 'Have you ever wondered how musical instruments make their sounds?' Whereupon there was a chorus of a hundred Welsh voices saying, 'Yes'. That was an important discovery—young children *answer* rhetorical questions and, if you do not want the answer, you should not ask! Furthermore, it soon became clear that children of these ages have none of the inhibitions about asking and answering questions that most of their older colleagues have. So I quickly replanned the style of presentation and conducted the rest of the lecture around questions to the children.

Since then, the age group 7–9 has become my favourite audience, and the question-and-answer technique my preferred method of presentation. Lectures to primary children at the RI began in 1985 and soon became so popular that the same lectures could be given four times on two successive days to completely full houses. At first they were confined to the 10- and 11-year-olds, but lectures designed specifically for 7-, 8- and 9-year-olds have now been presented for several years.

The questions-and-answer technique

The question-and-answer technique works well because it keeps the children involved from the word 'go'. It also fits in very well with the definition of science that I use with younger children. I tell them that science is just asking questions. When I do an experiment I am really saying to myself, 'If I did such and such a thing I wonder what would happen?'. Then I do it and find out. But, of course, sometimes I say to myself, 'If I did such and such a thing I know what would happen'. Then I do it and it doesn't happen. So the question then is, 'Why didn't it happen?'. And, of course, one often learns far more from that than from an experiment that works as expected.

I am a physicist and so most of the subject matter discussed here is physics, but the principles and techniques apply equally well to other sciences.

Simple aids

In addition to keeping the children involved by asking questions it is important to have something for them to look at. If I am asked to explain to children what sort of things physicists ask questions about I often use a handful of balloons, and it is quite surprising how much physics can be covered in this way.

The first question is, 'Why are they different colours?'. This is a cue to talk about the scattering and absorption of light and some simple ideas on colour vision. 'Why are the balloons flexible?' This leads straight into an excursion into some ideas about molecules. 'Why do the balloons stretch?' This is a question that leads to a discussion of long-chain molecules untangling. During this discussion I stretch a balloon and hold it for a few seconds to allow it to come to room temperature. I then place it on my top lip and let it relax while retaining contact with the lip. The balloon immediately becomes very cold. 'Why do I use the top lip?' (a) Because it is one of the most sensitive areas of the skin and (b) because it looks silly. A very useful rule in lecturing to this age group is that they will remember anything that is slightly silly, and experience shows that they remember the associated science. At this point one of the balloons can be inflated and the fact that the balloon is more-or-less round can lead to the simpler ideas of the kinetic theory of gases. If the neck is now stretched and some air released to make a squawk this could provide a lead-in to musical reeds. Finally, the balloon can be released to demonstrate conservation of momentum and rocket propulsion.

How a lecture proceeds

At this point in the discourse I showed a video recording of the beginning of one of my lectures to illustrate the question-and-answer technique. The topic was science and music and I began by asking 'What is sound?'. The first responses were surprisingly technical: 'Vibrations', 'Waves in the air', but eventually we arrived at 'Something that you hear'. The question 'What is pressure?' on this occasion produced quite good answers: 'Pressing down', 'Squash', etc. But sometimes the answers can be quite surprising, as for instance some time ago:

'It's like when daddy comes home and says he has had a hard day!'. I then use a bicycle pump and point out that the pressure changes involved in speaking are about the same as that produced by moving the handle of the pump about one-hundredth of a millimetre. Our ears are sensitive pressure gauges—but how do the pressure changes move from the source of sound to the ear?' Here we need to discuss waves and to emphasize that something moves, although the medium does not. I use a demonstration that was described by John Tyndall in his book *Sound*. Figure 1 is his illustration. Five children stand in line and we discuss what happens to the child at the front of the line if the child at the back is pushed. In this case the 'push' is transmitted but the children only sway about their positions.

Why younger age groups?

In the first place the enthusiasm of the 7–9 age group is boundless. They are like sponges ready to absorb all the information that is placed before them. Of course, it does not all stick, but a high proportion does. And the enthusiasm as they arrive for a lecture brings the whole building to life. I remember Lady Porter saying to me once that she loved days when the primary children were around because the whole building seemed to hum with excitement. During the discourse I played an audio recording of the deafening noise of the arrival of 400+ children! The effect on the children of entering the main theatre is quite striking; they immediately

Fig. 1 A 'boy wave' as illustrated in John Tyndall's book *Sound*.

look up at the dome and the comments are mostly variations on the word 'Cor!'.

Secondly, from a purely practical point of view it is much easier to gather together audiences in this age group because the school programmes are much less rigidly geared to timetables than those with older children.

And, of course, they have not yet made the curriculum choices that come as they approach GCSE and so they have a chance to see what science is like, not just as an examination subject.

It is important to remember that our aim is not to teach in the way that school teachers do. We aim to stimulate imagination and curiosity, and to give children the opportunity of seeing demonstrations that would be beyond the normal resources of schools.

Are there disadvantages?

Does presenting science in this way to such young children have disadvantages? Some people argue that I should not be quite so ready to accept the children's answers to questions, since clearly they often lack scientific accuracy. But I feel that up to the age of 9 or 10 it is much more important to expose children to the excitement of science and to the idea of questioning everything even if some dodgy ideas are acquired on the way. The same sort of argument holds for the use of scientific jargon. Some teachers have complained to me that they do not like my use of words like 'wobble' for vibration or 'squash' for pressure. But I see no objection as long as the idea is made clear and the scientific word is used and explained at some stage during the presentation. A friend of mine used to say that he believed that long scientific words should be introduced to 7- and 8-year-olds even though they have no idea what they mean. Having heard them at that age they will not be frightened of them when they meet them again later in life. His favourite illustrations were the two words 'energy' and 'entropy'. No one has any difficulty understanding 'energy' because it is a familiar word. But how many people have trouble with 'entropy' because they have never heard the word before?

Relationship to hands-on science centres

How does our lecture programme relate to hands-on science centres such as Techniques at Cardiff, which I pick out simply because I was involved in a small way in its original foundation? Obviously children love them and gain enormously from them. But I regard them as complementary to the demonstration lecture. They are, as it were, *à la carte* offerings: you can pick where you like and load up memory banks in a very fruitful way. But the lecture is *table d'hôte*—it takes you through topics in a logical sequence and I think that is a necessary complement to the hands-on scene.

Mixed ages

I am often asked how it is possible to deal with mixed age groups and mixed abilities. The solution again lies very much in the use of demonstrations. I usually plan to keep a lecture going at at least three levels simultaneously. I try to have something happening in the way of a demonstration at least every two minutes. Even the youngest or least able children enjoy watching something happen and, even if they do not understand it, they very often remember it and come back to it later. Then there is a middle level: the words that go with the demonstration together with the demonstration itself should keep the middle ranks interested. Then every now and again I throw out a titbit for the brightest children.

The significance of demonstrations

Demonstrations have many other good things in their favour and it would be out of place to spend too much time discussing them in great detail. Let me just make two points.

First, words can be misunderstood and if a lecture consisted solely of words it would be quite possible for sections of the audience to pick up erroneous ideas; but a demonstration that works is a positive gain, whether or not there are words to accompany it. An experiment that works is essentially truthful.

Secondly, demonstrations fix things in the mind. Sir Lawrence Bragg in the RI booklet 'Advice to Lecturers' says:

> *One is struck again and again by the immense superiority, as judged by the effect on the audience, of a series of experiments and demonstrations over a lecture illustrated by slides. The Christmas Lectures to Young People at the Royal Institution afford a good instance. It is surprising how often people in all walks of life own that their interest in science was first aroused by attending one of these courses when they were young, and in recalling their impressions; they almost invariably say, not 'we were told' but 'we were shown'.*

At this point in the discourse a short video recording of Sir Lawrence demonstrating Sir Richard Paget's models of the origin of vowel formants in speech was shown as an illustration of Sir Lawrence's brilliance in showing science to young people.

Getting the audience to help

It is always important to involve the children as much as possible. There are many ways in which this can be done apart from the question-and-answer technique. Sometimes a child without a specific skill can be used as assistant. An example of this may be taken from the 1989 Christmas lectures when a demonstration of resonance used a teddy bear sitting on a simple swing. The little girl who helped obviously enjoyed the experience so much that she made a memorable contribution. Another example from the same lecture involved a child who did have a specific skill—that of performing on the recorder.

Question sessions at the end

At the end of each lecture I always give the children the opportunity to ask me questions, and these sessions can go on for twenty minutes or more, often having to be cut short because of the arrival of buses. I have found an invaluable trick for starting questions off—which I have even successfully used with adult audiences. I tell the children that I shall give them 15 seconds during which they can relax the tension by making a bit of noise and then we will see who can get a hand up to ask the first question. A forest of hands goes up. Of course, some of the questions are trivial.

How old are you? How much do you earn? Have you ever blown anything up? But many have a serious content. I was once asked, 'When I dream about my favourite pop group are there sound waves in the bedroom'. Of course, the answer had to be, 'No, but your brain thinks there are'. The questions can be very stimulating and sometimes really put you on the spot. But sometimes they lead to a new slant on an old idea.

One of the experiments that is always a favourite with the children is Chladni's plate. I use it to begin to explain how using a sound box or board changes the quality of a sound as well as making it louder. This is one of the demonstrations that the children all remember, though I must admit that quite a few regard it as a conjuring trick and I have to spend time explaining that the patterns are characteristic of the plate—'All I am doing is finding the right one'.

On one occasion, when the demonstration was over, one child asked, 'Can you use anything other than sand?'. That started up a train of thought and I ended up reading Faraday's diary.

He asked the same question and went round the RI trying all the powders he could find. They all worked except one—lycopodium powder.

Figure 2 shows the Chladni plate on which the sand traces the nodal lines of a particular pattern and over which lycopodium powder has been sprinkled. When the plate is again excited by the bow the

Fig. 2 A Chladni plate which has been excited by bowing. The sand delineates the lines of least displacement and then lycopodium powder has been scattered on top of the sand pattern.

Fig. 3 The plate as in Fig. 2 is re-excited by bowing and the lycopodium now collects in the areas of maximum displacement.

lycopodium collects in the areas of maximum displacement, as shown in Fig. 3.

Faraday pointed out that the air above the areas of maximum displacement is thrown into turbulence, the powder is drawn up, and, when the bowing ceases, the powder falls back.

On a second occasion, I was playing the saw, which is another very popular experiment: it brings out the difference between very thin vibrators like strings, which cannot create large changes in the pressure of the air, and which need sound boxes to make the sound louder, and extensive vibrators that make enough sound on their own. After one such demonstration came the question, 'Could you put sand on the saw?'. We immediately tried the experiment and the results led to a much better understanding of the way in which the saw works. Figures 4 and 5 show the method of playing the saw and Fig. 6 is a drawing of a typical sand pattern.

Children's reactions

I always receive delightful letters afterwards, many of which include portraits of me and, surprisingly, a high proportion show me with a

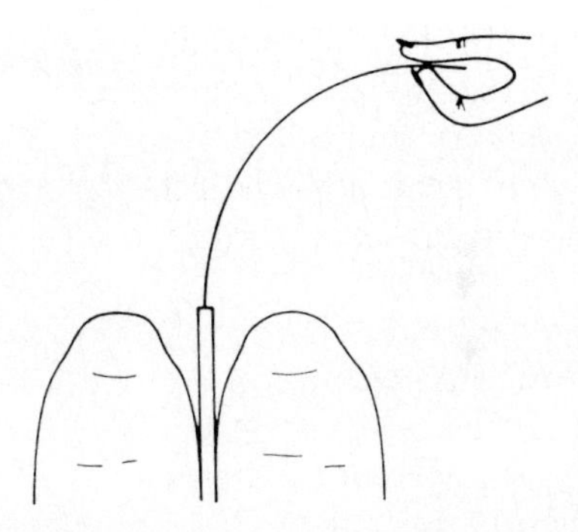

Fig. 4 If the saw is held between the knees and bent into an arc of a circle bowing produces no response.

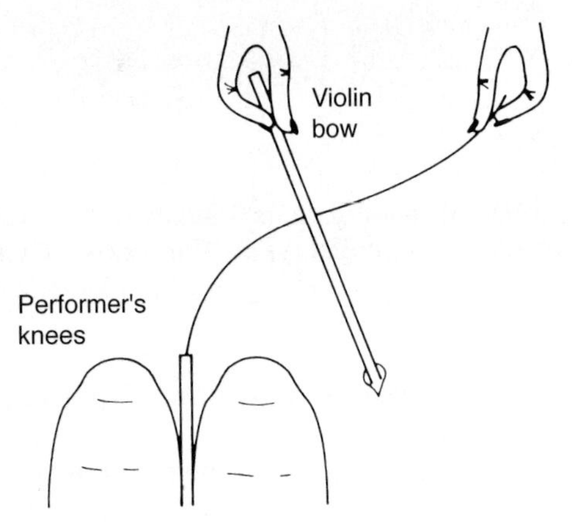

Fig. 5 If the saw is bent into an S-shape bowing produces an immediate singing response.

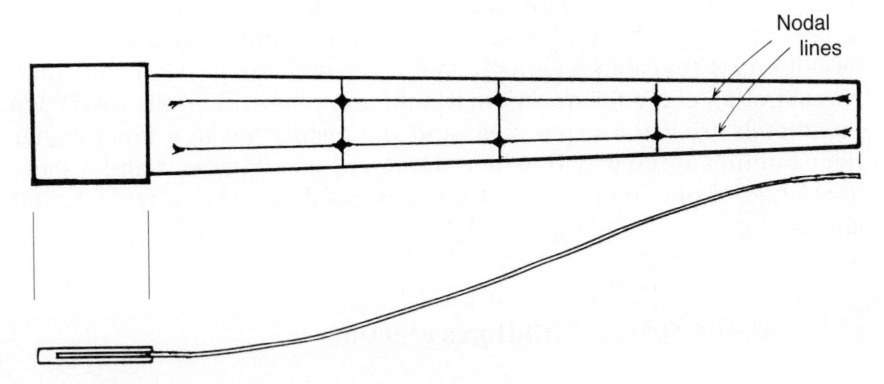

Fig. 6 A sketch of one of the sand patterns produced by a larger version of the saw, showing nodal lines.

beard! Presumably the children think that all good scientists should be bearded. Sometimes I receive confirmation that something really has been learnt. In that category one of my favourites followed a lecture on colour. It was a picture of a rainbow with the colours in the correct order together with the sun and raindrops which are, of course, the necessary requirements.

And this brings me to a final demonstration which always makes a big impression. In talking about colour and the spectrum I start with a small laser and point out that the beam itself is invisible unless the light falls on some object such as my hand. I then scatter flour in the beam and invariably the whole audience reacts with delight.

I then show how the laser beam can be deviated by a triangular prism. Finally, a slide projector with a narrow slit in the slide position and a red filter over the lens is used (Fig. 7) and the position of the beam deviated by a prism is marked on a white card. We then discuss what will happen if the colour is changed, first to green, and then to blue. Usually about half the audience vote for the right answer with green, and then a higher proportion with blue, obviously having worked it out. But white light seems to throw them completely and hardly any children in these age groups seem to be able to predict the full spectrum. The effect of removing the blue filter to show the full spectrum is very dramatic and the sound of 400 ‘oohs’ and ‘aahs’ is one of the many rewards of this kind of presentation.

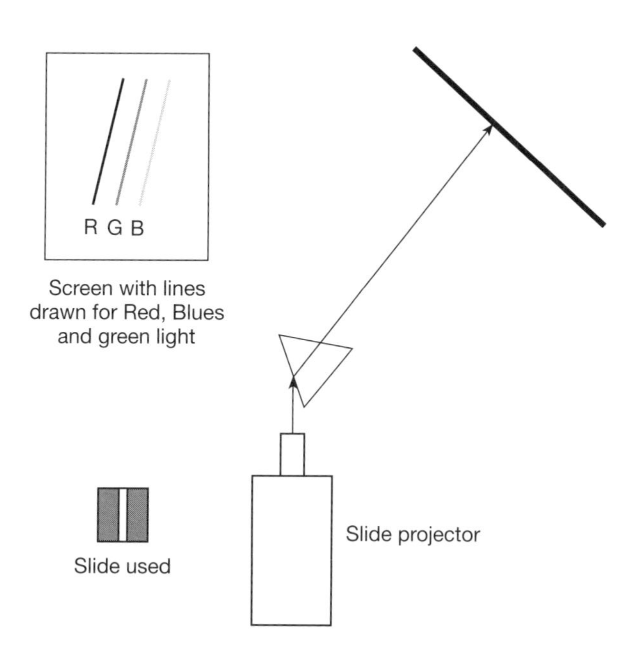

Fig. 7 The projector arrangement for producing a spectrum.

Acknowledgements

I am very grateful to professors Ronald King, David Phillips, and Richard Catlow, who successively invited me to give the school lectures on which this discourse is based and to Mrs Jean Conisbee for making the associated administrative arrangements.

CHARLES TAYLOR

Charles Taylor was Professor of Physics at University College, Cardiff from 1965 to 1983, and from 1977 to 1989 he was also Professor of experimental physics at the Royal Institution. He gave the televised Royal Institution Christmas lecture in 1971 (Sounds of music) and in 1989 (Exploring music) and in 1986 he became the first holder of the Royal Society's Michael Faraday Award for contributions to the public understanding of science. The most relevant of his publications to this discourse are *The Oxford children's book of science* and *The art and science of lecture demonstration.*

The Mayfair properties of the Royal Institution

H. J. V. TYRRELL

The name of the Royal Institution is surely associated in the public mind with that vision of the Albemarle Street frontage of No. 21 with the fourteen three-quarter-round Corinthian columns rising up to the third-floor level. Although these are part of an architectural fancy, entirely devoid of structural significance, and applied in 1836 to an eighteenth-century façade to enhance the importance of the Institution in the eye of the beholder, the success of the concept in achieving that aim is undeniable. It is less well-known that the Institution is now a substantial property holder in Mayfair, holding headleases on property on the east side of Albemarle Street northward from No. 21 up to Grafton Street, on the south side of Grafton Street from Albemarle to New Bond Street, and on 165–167 New Bond Street. In addition, it owns the freeholds of 18, 19, and 20 Albemarle Street. All the leases are from the Corporation of the City of London and run from 29 September 1921 for 2000 years at rents which are essentially the same as those fixed for earlier leases in the eighteenth century. At present all the activities of the Institution are confined to 20 and 21 Albemarle Street although until very recently some of the accommodation in No. 19 was also used. However, with the exception of 22 Albemarle Street and 166 New Bond Street, which are let on term leases at modern rental values, the other leasehold properties are underlet on 2000-year leases from 1921 at low fixed rents and are of little financial value to the headleaseholder. Equally, the value of all the headleases to the freeholder is also small. To understand this curious situation it is not only necessary to examine the history of property acquisition in the Royal Institution itself but also to consider how the ownership of the land on which the buildings stand has changed since the early seventeenth century.

The early history[1]

The roads now known as Piccadilly and Oxford Street preserve the lines of the two ancient routes leading westward from medieval London. Prior to the Restoration of Charles II the area between these two roads retained its rural character. Immediately north of the line of Piccadilly there was land belonging to the Crown but, further north and to the east of the Tyburn brook a large area of land stretching up to the line of Oxford Street had been transferred from the Crown to the Corporation of the City of London in 1628 in payment of an earlier debt. This area became known to the Corporation as the Conduit Meads Estate and was once of importance as a gathering ground for the supply of water to the City, see Fig. 1. This function was the origin of the name of Conduit Street, and a pump house, whose position is noted in the early leases, stood close to the present-day junction of Bruton Street with New Bond Street. In this locality the line of the Tyburn, also known as the Ay-Brook, is marked approximately by the line of the modern Bruton Lane at its upper and lower ends.

Edward Hyde, Earl of Clarendon, had been an adviser to and companion of the future Charles II during his exile and became Lord Chancellor after the Restoration. In 1664, he and his son Henry, Viscount

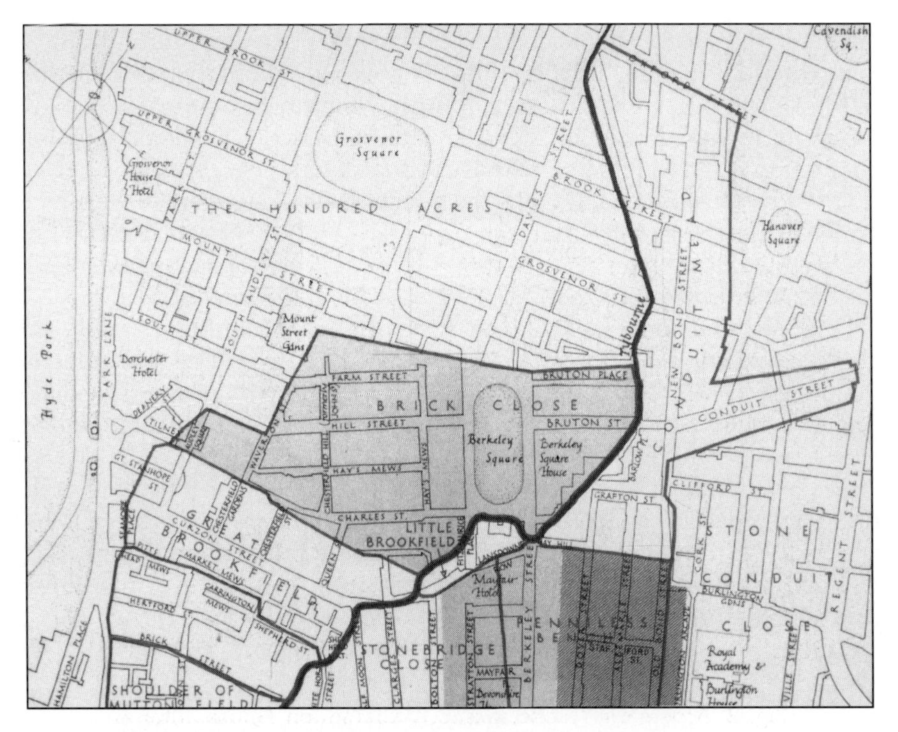

Fig. 1 The approximate medieval field boundaries superimposed upon the modern street pattern in Mayfair. Reproduced from an end-paper in Reference 1.

Cornburg, were rewarded by the grant of the freehold of 29.5 acres of Crown land in a block which extended north from Piccadilly (then called Portugal Street in honour of Charles's queen, Catherine of Braganza) approximately to the present line of Hay Hill. The western boundary was the line of the Tyburn and it stretched eastwards to beyond the site of the present Burlington House. They immediately sold about 13 acres in two separate blocks on the western side and one of about 8 acres to the east. They retained the central block (the heavily hatched area in Fig. 1) of about 8 1/2 acres stretching, in modern terms, from the backs of the houses on the western side of Dover Street to the line of Burlington Arcade up to Burlington Gardens and thereafter along Bond Street, stretching northwards up to the boundary of Conduit Mead. Edward Hyde there built (1664–7) Clarendon House, designed by Roger Pratt as a great country house and described by Pepys as 'the finest pile I ever did see', both for his residence and for his official duties (Fig. 2). With the intention of providing a park fit for such a house Clarendon obtained a 99-year lease from 1668 on 24 acres of the Conduit Meads land, intending to have it laid out by John Evelyn. That intention was

Fig. 2 Albemarle House, formerly Clarendon House, after an engraving (1682) by William Skillman of a drawing by John Spilberg. Reproduced by permission from the Crace Collection, British Museum.

probably not carried out and the house was never completely finished although Clarendon lived there briefly during 1667. Its great cost (£50 000 according to Evelyn) and popular suspicions about the source of the money, coupled with Clarendon's unpopular policies, led to his downfall and exile to France where he was able to complete his monumental *History of the Great Rebellion* before his death in 1674. His son then sold the whole of the freehold and leasehold estate for £25 000 to the second Duke of Albemarle who renamed the house as Albemarle House but in 1683 sold it to a syndicate headed by John Hinde, a City goldsmith, for £35 000. The house was demolished in that year and a partnership of Sir Thomas Bond, a financier with considerable influence at Court, Henry Jermyn, shortly (1685) to be created Baron Dover, and a wealthy lady named Margaret Stafford, agreed to take up building leases on large parts of the estate known by then as the Albemarle Grounds. Dover, Albemarle, Stafford and Bond Streets began to be laid out, Dover Street and Bond Street appearing in the rate books as early as 1686 with Albemarle Street some ten years later (Fig. 3). The original plan included the construction of a large square on the leasehold land north of the freehold section of Dover Street, Albemarle Street, and the western side of Bond Street. The only part of this plan to be realized was the creation, on the leased land, of an east–west road between Bond Street and Hay Hill to form the southern boundary of the planned Albemarle Square. Thomas Bond died in 1685 and Hinde was made bankrupt in 1687, events followed by numerous law suits in the Chancery Court. Development was therefore very slow even on the freehold land close to Piccadilly.

In 1694 the City of London assigned the residue of the Clarendon lease on about 4 acres of Conduit Meads land to the Hon. John Sheffield, third Baron Mulgrave. Sheffield was a skilful politician who had been a member of the Privy Council under James II, lost that position when James fled abroad, yet found sufficient favour under the new regime to be reappointed to the Privy Council, awarded a pension of £3000 per annum, and created the 1st Marquess of Normanby, all in 1694, clearly a good year for him. His political power at that time may have been the reason why the City of London allowed him to secure an extension of the lease by 100 years to 1867, an act of generosity which later cost the City dear. Figure 3 is based on a plan attached to this lease and shows the plot to to have been bounded on the west by the line of the Ay-Brook (the Tyburn), on the north by a line parallel to and slightly north of the short stretch of Bruton Street between its junction with Bruton Lane and New Bond Street, on the east by New Bond Street itself, and on the south by the northern boundary of the freehold land which had been assigned to Clarendon. Within a year Sheffield had assigned the remainder of the 99-year term of this Clarendon lease on his newly acquired land to John,

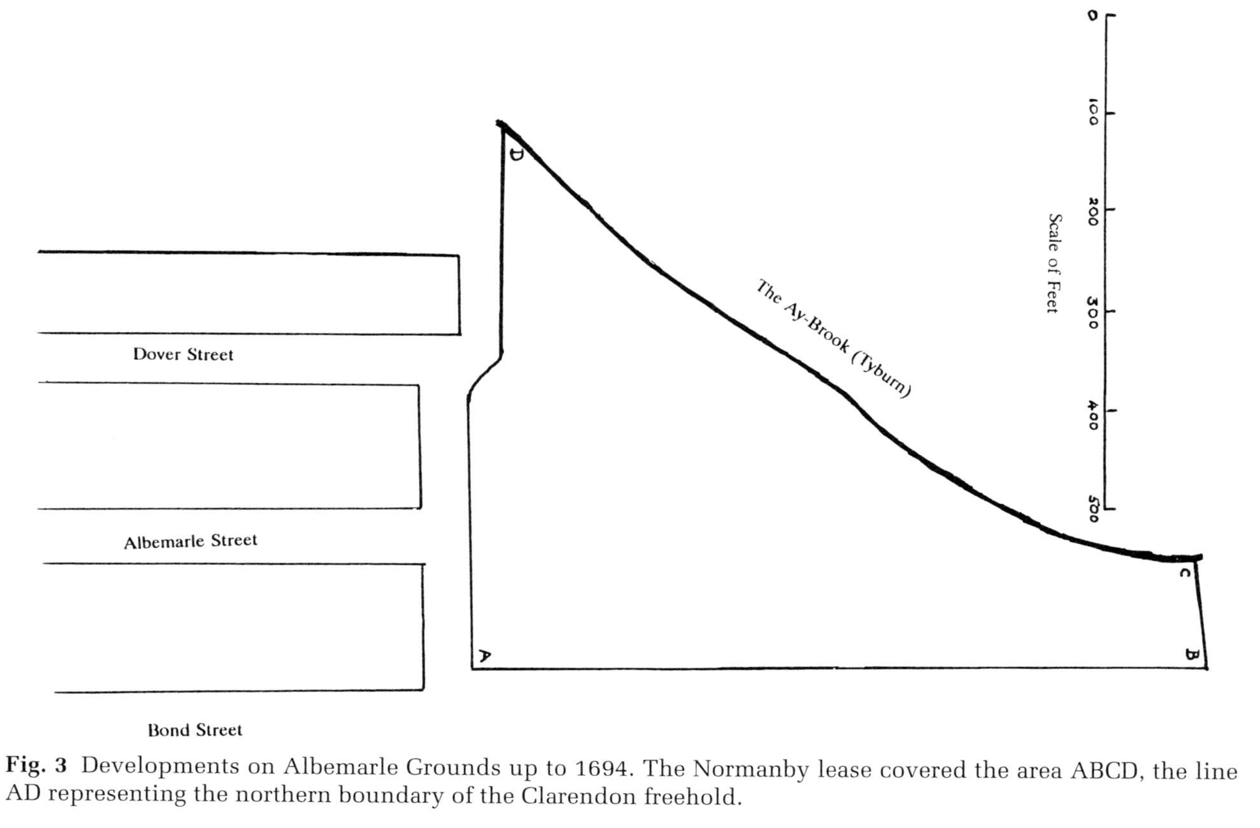

Fig. 3 Developments on Albemarle Grounds up to 1694. The Normanby lease covered the area ABCD, the line AD representing the northern boundary of the Clarendon freehold.

Lord Jeffreys, but reserved the extension from 1767 to 1867, granted to him by the City of London, to his own heirs and successors. In Fig. 3 the blocks representing Dover, Albemarle, and Bond Streets terminate about forty feet south of the lease boundary, and, both from this and the plan shown in Fig. 4, it seems that there was in 1694 an older road of this width, on freehold land, running between Bond Street and Dover Street and then turning slightly north to join with Hay Hill. This was the road referred to in Fig. 5 as the 'Old Road', later replaced by the parallel road on the leasehold land intended to form the south side of the proposed Albemarle Square. Sections of the 'Old Road' seem to have been earmarked for freehold building plots and two of these lay between Bond Street and Albemarle Street each with frontages of about 65 feet to the new road, the freeholder then being Lord Dover. The plot on the corner, with a frontage of about 40 feet to Albemarle Street, was sold by Dover in 1703 for 21 years' purchase of the ground rent (13 guineas per annum) to Robert Fryth, a plasterer, and Richard Fryth, a bricklayer. They built the house and sold it in January 1705 to the trustees of Henry d'Auverquerque, first Earl of Grantham, for £3000. The plot on the corner of Bond Street, with a frontage of 51 feet to Bond Street, was bought, also in 1703, by Thomas Barlow, a carpenter, who built the house and sold it to the second Duke of Grafton, a grandson of Charles II and Barbara Villiers. Jeffreys died in 1702 leaving a widow and an infant daughter and, in 1707, his executors re-assigned his lease on the land shown in Fig. 3 to a syndicate consisting of Grafton, Grantham, Viscount Hinton St George (another resident of Albemarle Street), and Alicia Wallop, widow. The partners shared the land among themselves, Grantham and Grafton obtaining blocks stretching north from their respective freeholds. Either then, or later, Grafton obtained leases on additional land, as can be seen from a plan, reproduced in Fig. 4, attached to a City lease of 1767 to the third Duke.

The new leaseholders obtained a writ of '*ad quod damnum*' in 1709 which permitted them to move the then existing east–west road just north of their freeholds in Dover, Albemarle, and Bond Streets about 90 feet further to the north, allowing Dover and Albemarle Streets to be extended in the same direction. A later writ, apparently in 1723, led to the 1709 east–west road being taken up and replaced by one on the site of the present Grafton Street, thus establishing the existing street pattern. This staged development is shown in Fig. 5, redrawn from the City Comptroller's Lands Plan No. 386A, compiled after 1723 and before Grantham's death in 1754. The 'Old Road' marked thereon evidently lay on freehold land, and the changes from 1709 onwards gave Grantham, Grafton, and the other freeholders named in Fig. 5, blocks of leased land running north to Grafton Street from their respective freeholds. The Figure shows a building on Grantham's lease with a frontage, scaling at

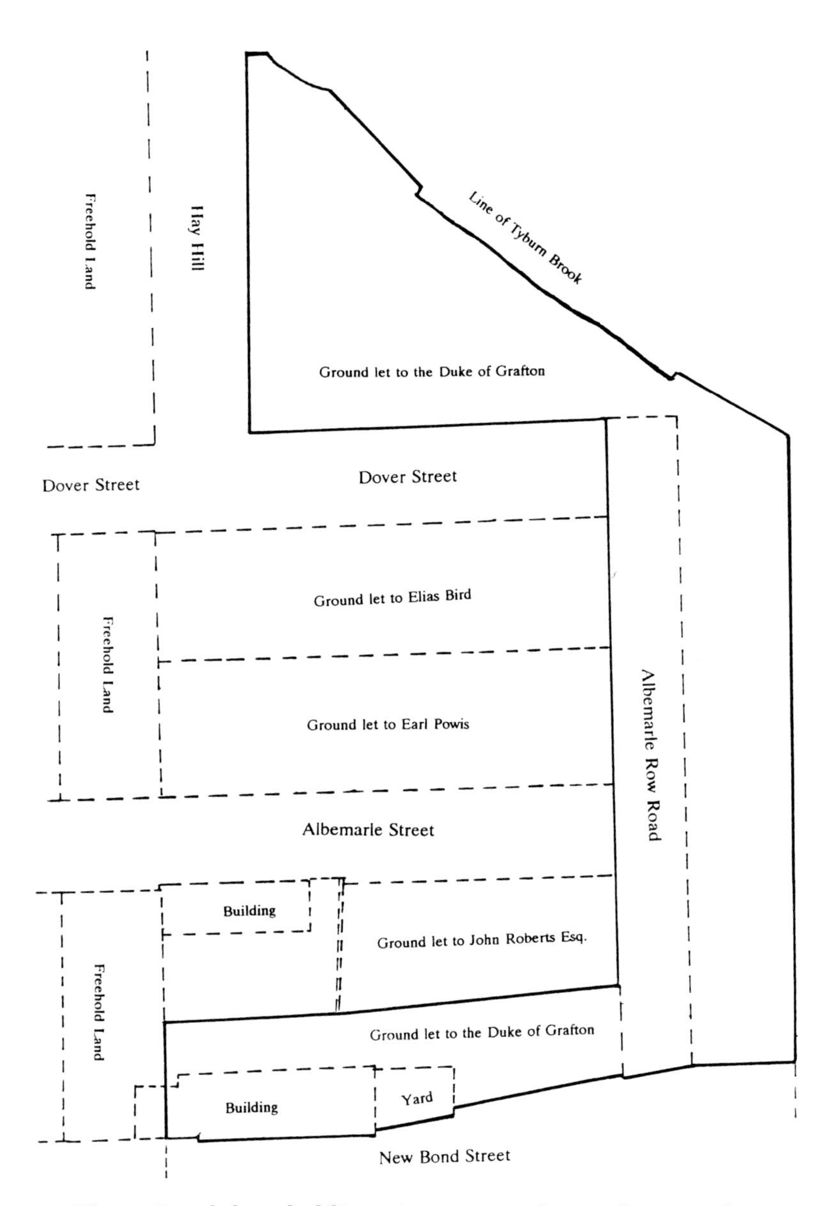

Fig. 4 Local leaseholdings in 1767, redrawn from a plan attached to a City lease of that date granted to the third Duke of Grafton.

about 76 feet, to Albemarle Street, apparently constructed between the dates of the two writs. This appears on Rocque's map of 1746 and is identifiable as the original part of 21 Albemarle Street. Its position in

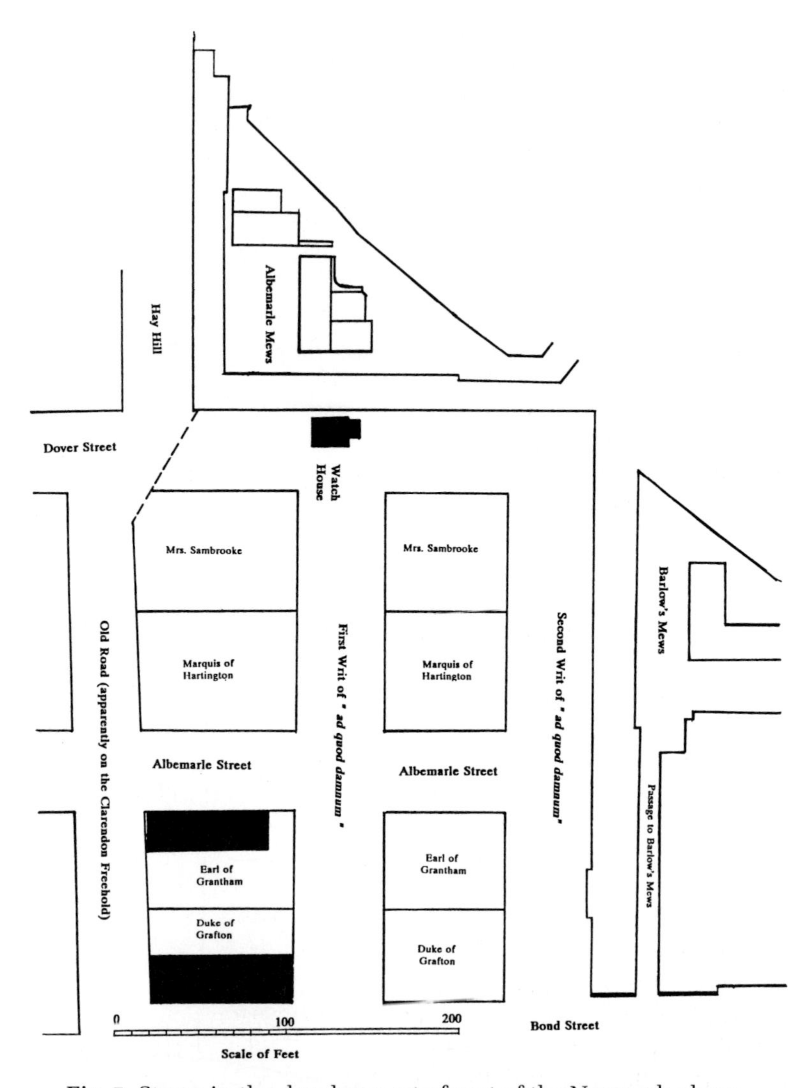

Fig. 5 Stages in the development of part of the Normanby lease. Dark areas represent some of the buildings existing before 1755. Redrawn from Comptroller's City Lands Plan No. 386A.

Fig. 5 confirms the view that the 'Old Road' was sited on Clarendon's freehold.

The choice of 1723 for the second writ is based on a report in the *British Journal* for 30 March in that year, which stated that 'All the waste ground at the upper end of Albemarle Street and Dover Street is purchased by the Duke of Grafton and the Earl of Grantham for gardening and the road there, leading to May Fair, is ordered to be turned'.[2] Like many press reports this one was not entirely accurate but it is strong evidence for the date of the second writ. Since the Earl of Grantham and the Duke of Grafton are named in Fig. 5 as the holders of the leases on the east side of Albemarle Street and the west side of New Bond Street and Mr John Roberts bought the freehold of 20 Albemarle Street together with the leasehold land to the north in 1755 after Henry Grantham's death at the end of 1754, the plan in Fig. 5 must precede this date. A slightly later City of London plan (Fig. 6), prepared after Roberts bought the freehold of No. 20 and the house was occupied by Mr Mellish, shows a ground plan of No. 21 in which the frontage on Albemarle Street is marked as precisely 76 feet and marks what seems to have been the mid-line of the 'Albemarle Square' road about 25 feet north of the freehold–leasehold boundary. At about this time the north side of Grafton Street (called 'Evan's Row' by Rocque and 'Albemarle Row Road' in Fig. 5) was occupied by stables, as was the west side of the extension of Dover Street beyond Hay Hill (now Grafton Street) with a watch house set in the street itself and mews behind the stables.

Expiry of the Clarendon lease

The 99-year lease from the City of London to the Earl of Clarendon was due to expire in 1767 but the successors in title to the first Marquis of Normanby would have retained control over the four-acre plot obtained by Normanby in 1694 by virtue of the 100-year extension which had been negotiated at that time. All other leases, such as those held by the third Duke of Grafton and John Roberts were due to expire in 1767 and as that year approached those leaseholders needed to negotiate new leases if they were to retain title to their property in the area. Also the Court of Common Council of the City became aware that control of what was, by that time, a valuable property would be denied to them until 1867 unless they could buy back that lease extension.

Normanby, who had been created the first Duke of Buckingham and Normanby in 1703–4, had died in 1721 and was succeeded by his only surviving son. The second Duke died without issue in 1735 at which point the titles became extinct, the lease extension option passing with other property to his heirs. The Court of Common Council eventually negotiated a price of £24 000 for the lease to be re-assigned from those heirs to the Trustees of the City Corporation. This was a very large sum for the City to find in cash and the Minutes of the Court of Common Council of the time reflect years of argument about the financing of this before the matter was finally settled in August 1765. Two years later, the City offered to grant leases of a new kind to existing leaseholders on the Conduit Meads estate. These were leases of either 40 or 61 years (less one quarter) which could be renewed in perpetuity by payment of a

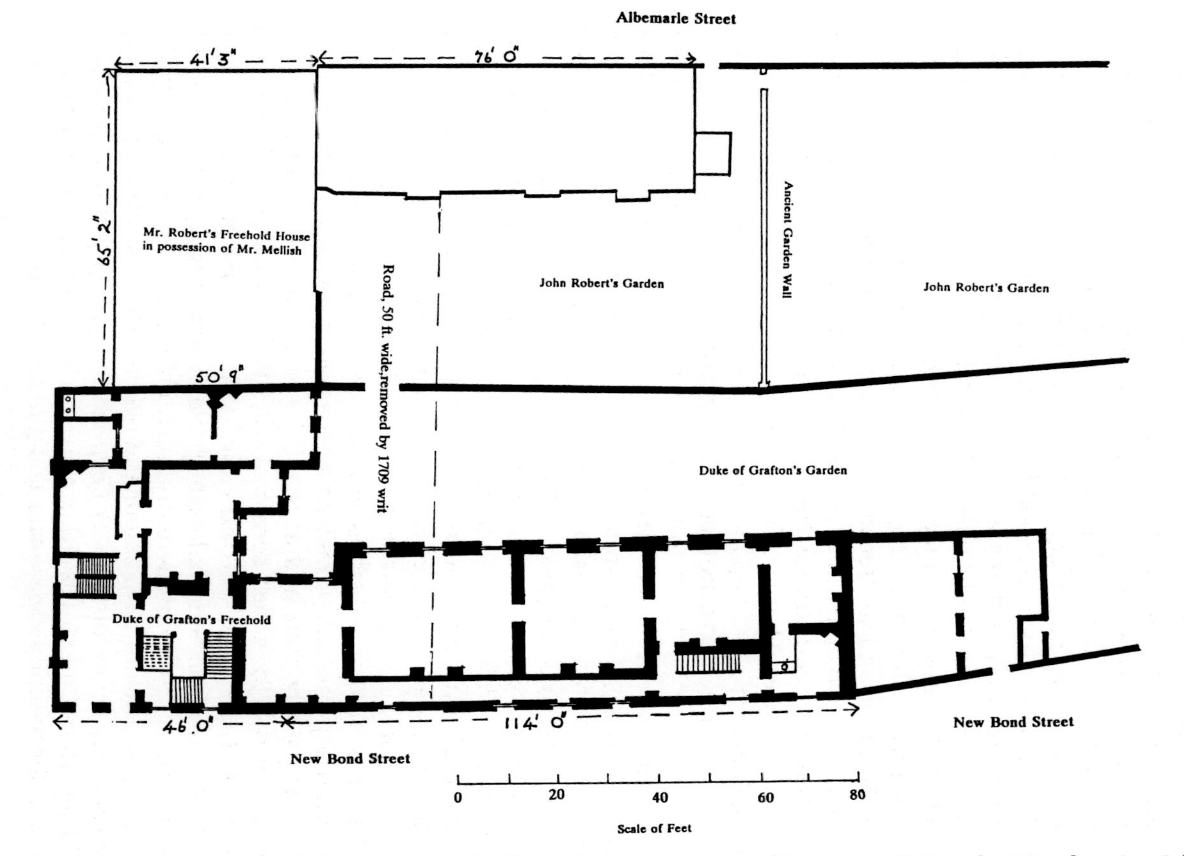

Fig. 6 Redrawn from Comptroller's City Lands Plan No. 190A, prepared between 1758 and 1772, showing John Roberts to have owned 20 and 21 Albemarle Street with William Mellish as his tenant in No. 20.

lease renewal fine for every fourteen years. If these fines were not paid by the due date the freeholder could refuse to accept that fine or any further fines. In such a case the lease would terminate after the agreed term had elapsed from the payment of the last fine. John Roberts was granted a lease of 61 years less one quarter on 21 Albemarle Street and the associated open land up to Grafton Street for a ground rent of £150 p.a., subject to a lease renewal fine of £600 due in 1767 and at fourteen-yearly intervals thereafter. The third Duke of Grafton obtained a similar type of lease on the much larger plot shown in Fig. 4. That rental was set at £455 p.a. with a lease renewal fine of £2200. There was no provision for regular reviews of these sums. By 1795 a plan (Fig. 7) attached to another lease on this land obtained from the City by the Duke shows it to have been almost entirely redeveloped with houses and offices, some of which survive to this day along the west side of Grafton Street.

The Westminster Rate Books appear to show that John Roberts occupied 20 Albemarle Street in 1756 and 1757 and that William Mellish was responsible for the rates on No. 21 in those years. Thereafter Roberts paid rates on No. 21 up to his death in 1772 and Mellish took that responsibility for No. 20 up to 1775 when he moved into No. 21.[3] At that time that house had a frontage of only 76 feet to Albemarle Street, extending from No. 20 up to the north wall of the well of the present Grand Staircase. When the house came into the possession of the Royal Institution in 1799 it had been extended northward by over 30 feet, possibly by John Roberts or, more probably,[3] by William Mellish. When John Roberts's lease was assigned to him by John Roberts's heirs in 1775 he apparently had to agree to take over responsibility for building and other contracts into which John Roberts had entered before his death. During William Mellish's tenure he granted separate underleases on two building plots on the open area north of No. 21 up to Grafton Street. The first of these, dated 2 November 1776 to a Mr Edward Gray , ran from Michaelmas 1767 for 61 years less one quarter at a ground rent of £74 p.a. and was perpetually renewable provided that a lease renewal fine of £370 was paid in 1781 and at fourteen-yearly intervals thereafter. This plot was at the northwest corner of Albemarle and Grafton Streets and was where Gray built 23 Albemarle Street. This house survived until about 1896 when it was replaced by the present complex of shops and apartments. A similar perpetually renewable underlease was granted in July 1782, to run from Michaelmas 1781, to a Mr Swinton for the land between Gray's plot and No. 21 and was where Swinton built No. 22, a building which, much altered, survives to this day as part of Asprey's. The rent for Swinton's lease was £48 p.a. and the lease renewal fine was set at £240.

Both Gray and Swinton built these houses as investments to provide rental income. Gray complicated the lease pattern at some time after

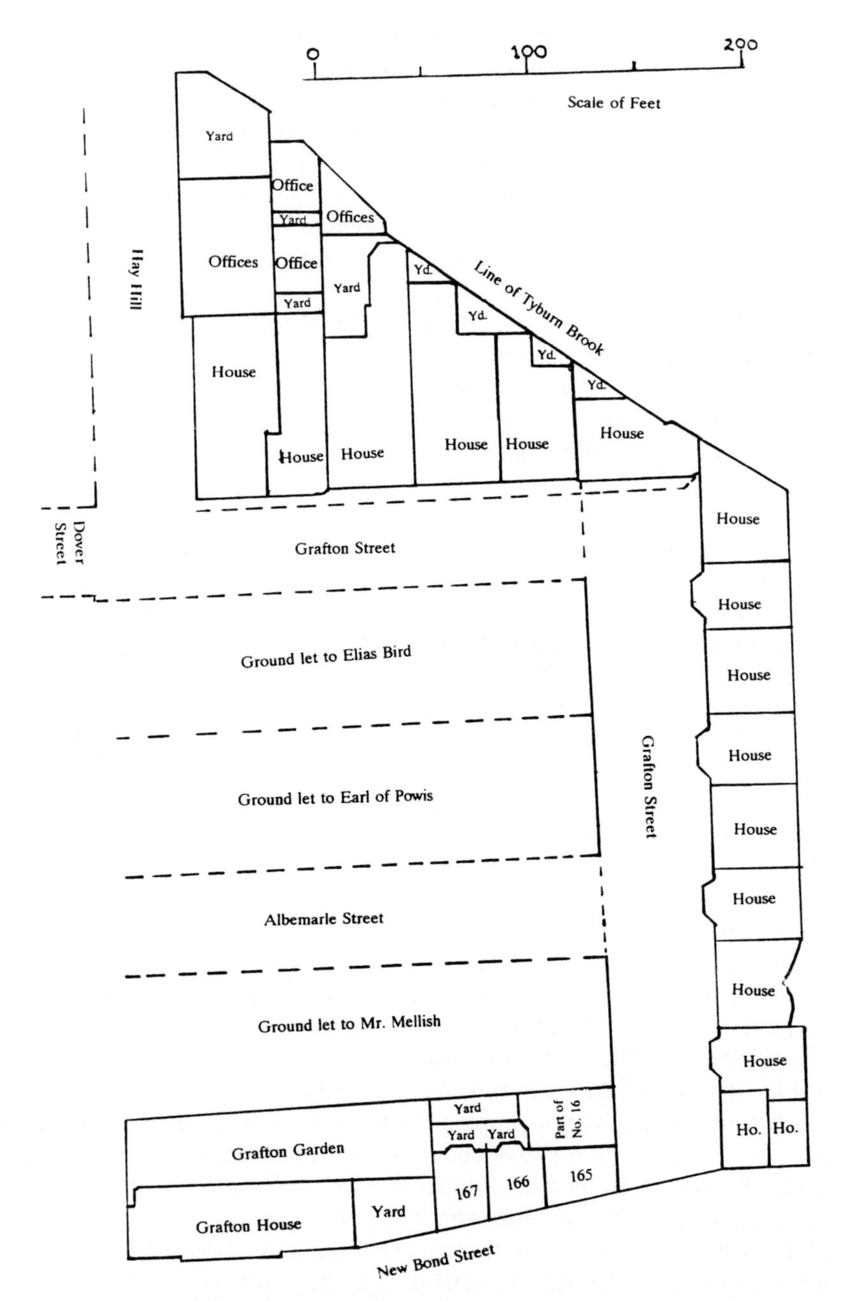

Fig. 7 Building development in Grafton Street area by 1795. Redrawn from plan attached to the 1795 renewal lease to the third Duke of Grafton.

1776 by subleasing a narrow strip from the east side of the land he held from Mellish. This ran south from the Grafton Street boundary as far as the north wall of No. 22 and can still be seen on the Land Registry map of the area (and also in Fig. 8 on page 214). It forms part of the site of 16 Grafton Street the major part of which was built on land leased from the City by Grafton. The sublease was to a Mr Paynter, whose family held it until 1919 when it passed to Asprey and Co., at an annual rent of £4 p.a. and subject to a lease renewal fine of £20 at the usual fourteen-year intervals.

William Mellish was clearly a good business man. He had acquired the headlease of No. 21 and its undeveloped land cheaply and, after granting leases to Gray and Swinton, paid a net ground rent to the City of £28 p.a. and could expect to receive £10 more in lease renewal fines from his tenants than he himself would have to pay to the City. For that small outlay he gained a large house in a fashionable residential area.

A house for the Royal Institution

Within a short time from the initial meeting early in 1799 to discuss the foundation of the Royal Institution Count Rumford and Thomas Bernard were asked to look for a home for the nascent Institution. Initially they considered the purchase or lease of a house in Bond Street formerly occupied by the recently deceased Lord Buckinghamshire. His executors were prepared to accept a rental of 400 guineas p.a. with an option to purchase for £4500. When coupled with rates of £128 p.a. this was not regarded by the Managers as affordable 'in the present state of the Funds unless the freehold part of the land at the south end of the site could be let off or disposed of at a fair rent'. Accordingly Rumford and Bernard were instructed to make further enquires. These led to the suggestion[4] that the late Mr (John) Mellish's house in Albemarle Street 'might prove a very proper situation for the Royal Institution'. William Mellish's name disappeared from the Westminster Rate Books after 1792, to be replaced by Anne Mellish (1793–5), and then by John Mellish (1796–9). John Mellish, the son of William had been killed in 1799 in a scuffle with a highwayman and, in consequence, his executors wished to sell his residence, No. 21 at a price expected to be between four and five thousand pounds. Henry Holland and John Soane agreed to carry out a survey, without charging a fee, and reported favourably on its condition. William Mellish's policy of developing the site meant that the net ground rent was low, a small profit on the lease renewal fines was to be expected, and, furthermore, the rates were much lower at £83 15s than those on the Bond Street house. Holland and Soane valued the house at £4500, including fixtures, and Bernard was authorized to negotiate a purchase at this figure. By 20 April he was able to report that he had agreed a figure of £4500 for the house but that fixtures, believed to be

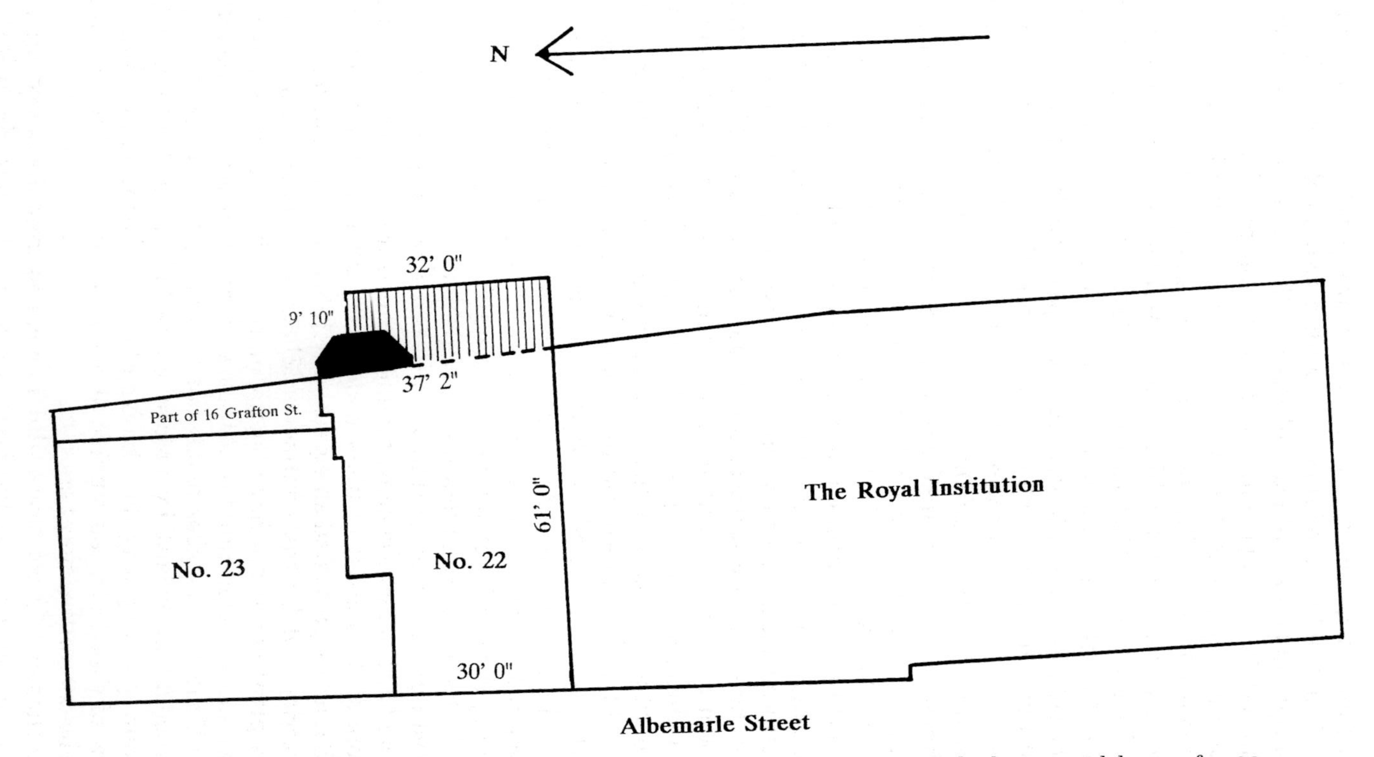

Fig. 8 Redrawn from William Pole's report (1862) of problems in connection with the ground leases for 22 Albemarle Street. Black and hatched areas show land leased by Baldry to Spooner in 1828.

worth about £200, were to be paid for by the purchaser. The vendors would clear the house of furniture within two weeks but wished to retain a right to an unpaid lease of £240, with accrued interest from the due date (1795) of payment. Clearly Swinton had failed to pay his renewal fine. Although the purchase was not finally agreed until 14 September 1799, the first meeting of the Managers in the new premises was held on June 5th while final terms were still being discussed. Eventually a sum of £4850 was agreed for the whole of the Mellish lease including No. 21, the fixtures remaining therein, and all the leasehold interests; subsequent events showed that these included all unpaid lease renewal fines though this is not mentioned explicitly in the Managers' Minutes. The deed of conveyance was dated 29 August 1799 but the seal of the Institution was not affixed until 9 February 1801.

The house naturally required adaptation for the public purposes of the Royal Institution and Caroe's book[3] gives a detailed account of the process of change. A Mr Spiller, a surveyor and life subscriber, was employed[5] to do an accurate measurement survey, and to collaborate with Rumford in preparing sketches and designs for the necessary changes to be submitted to a newly established Buildings Committee. Work on the lecture theatre seems to have begun in April 1800 with the demolition of part of the Mellish northern extension of the house in order to create sufficient space for the planned theatre which was to extend northward over what had been a small garden up to the boundary with No. 22. Spiller presented his report with plans and elevations on April 21st and was asked to prepare cost estimates. However, at the meeting of Managers on May 12th it was reported that Mr George Saunders of 252 Oxford Street had offered his 'gratuitous services' in conducting the works about to be executed. This offer was much to Rumford's taste and Saunders was promptly named as Surveyor to the Royal Institution, following Holland and Soane who had held the office jointly, and Spiller, whose name then disappears from the records except for a note of the receipt of his account for £120 15s.[6] The speed with which Saunders presented an agreement on May 26th for building work to be carried out by the contractor, Thomas Hancock, suggests that Spiller's plans could not have been greatly modified. Later minutes in that year show the payment of stage payments to Hancock, an offer to supply all the copper tubes required for warming the lecture theatre to be supplied 'at prime cost' by a Mr John Grenfell, and a clue to the original colour of the seating therein. Rumford was requested 'to provide cushions of Green Moreen stuffed with flocks', for all the seats, which were to be continuous benches.

These changes did not pass without comment from the neighbours. In 1800 No. 20 was occupied by the Dowager Lady Lilford, the widow of Thomas Powys M. P. who had been living at that address in 1796, was created Baron Lilford in 1797, died in 1800, and was succeeded by his son. Lady Lilford was 'apprehensive about the inconvenience' and the Managers asked the Building Committee to discuss matters with her surveyor. The affair was eventually settled amicably enough by the provision[7] of annual tickets of admission to the lectures for Lady Lilford and her three daughters in recognition of the inconvenience, 'to which the neighbourhood of the Royal Institution had subjected her Ladyship'. A more formal objection came, through his solicitor, from the Duke of Grafton, apparently on the grounds that the new buildings on the site of No. 21 had approached too closely to his premises in New Bond Street and he was, therefore, reluctant to accept a draft of Articles of Agreement intended to resolve the dispute. There was a somewhat tart reply from the Managers to the effect that the disputed buildings were entirely within the party wall and entirely upon their own premises.[8] That seems to have settled the matter since there is no further reference to it in the Minutes.

Even in 1800 the Institution was faced with the task of building on a restricted site in a residential district which was almost fully developed. There was a need for an open site close by in order to 'deposit materials'. Fortunately there was such a site on land then leased by Lord Suffield on the opposite side of Albemarle Street where No. 27 now stands, and Suffield was generous enough to allow the Managers[9] to use the space free of charge, subject only to clearing the site when it was needed for development. Later lease plans, and a reference in the Minutes in 1811,[10] show that, within a few years, a large private chapel (St George's Chapel) occupied the site until replaced by the present building (No. 27) in the early years of the twentieth century.

Problems with the Gray and Swinton leases

When the Institution bought the headlease on 21–23 Albemarle Street in 1800 the lease renewal fine of £600 due in 1795 had been paid by the Mellish family, thus securing the possibility of lease renewals in perpetuity. The next fine was due in 1809, and it was expected that the fines on the underleases to Gray and Swinton would more than recompense the Institution for this. Matters were not quite so simple.

Mr Edward Gray had died early in the first decade of the nineteenth century and the Court of Chancery had appointed a Receiver, Mr James White, for his estate. In the early summer of 1807 White wrote to the Managers on the subject of the renewal of the underlease for the corner plot on the northwest corner of Albemarle and Grafton Streets on which Gray had built 23 Albemarle Street as an investment. It was clearly a good address. The ninth Earl of Thanet was living there in 1796 and in

1802,[11] presumably with the Hungarian lady whom he was said to have 'carried off from her husband in Vienna' and later (1811) married in St George's, Hanover Square, when the marriage documents described her as a widow.[12] The tenant in 1808 was the Rt Hon. Sir John Anstruther, Bt, a former Chief Justice of Bengal who had been made a Privy Councillor in 1806. Mr White disclosed that Gray had failed to pay the lease renewal fines on his underlease when they were due in 1781 and in 1795, and he wished to know whether the Managers would accept payment of these fines, together with interest at 5% accumulated on the debt, from funds in Gray's estate and thereby reinstate the perpetually renewable nature of the underlease. It seems that the Managers had previously been unaware of this situation since the only reference to unpaid lease renewal fines in their minutes was that due in 1795 for Swinton's lease on No. 22. The decision was an important one since, in the strictness of the law, Gray's right to renewal had been lost after 1781 and his underlease would have reverted to the Royal Institution at Midsummer 1828. But this unexpected debt to the Institution stood at £1443 in November 1807, and a further £370 was due in 1809 when the Institution would itself be faced with the need to pay £600 to the City of London in respect of its own lease renewal fine. Thomas Bernard was asked to discuss the matter with the Court Receiver and with Mr Morgan, the solicitor to the Institution. A draft report to the Proprietors, prepared by Bernard and Morgan, was considered by the Managers.[13] This also disclosed that the Mr Paynter, to whom Gray had granted a sublease on part of his own leased land, also held an additional perpetually renewable underlease from the Duke of Grafton on an adjoining plot. This was also in danger because of unpaid lease renewal fines. Paynter, like Gray, had died by this time and his estate was also under administration in the Court of Chancery. The Receiver had made an application to the Duke for lease renewal in terms similar to those offered to the Institution, and that application had already been accepted. That fact, and the acute shortage of funds within the Royal Institution at the time, persuaded the Managers and the Visitors to recommend that the Proprietors should approve the Receiver's proposal and accept the amounts due. Early in 1808[14] that approval had been obtained and Mr White was informed. No answer had come from the Receiver by the end of May 1808 and, on 13 June, Thomas Bernard reported that he had written to the solicitors dealing with Gray's estate concerning their failure to comment on the terms of a draft lease which had been sent to them earlier. They had replied to the effect that they intended to proceed with the scrutiny of the lease but no data could be given for this since the partner dealing with the matter was 'out of town'. This behaviour angered the Managers and they therefore instructed the Institution's solicitor to take back the draft lease and not to proceed further with its renewal. Hearing of the situation, Sir John Anstruther, then the tenant of No. 23, offered to take over the lease himself by paying the outstanding fines with the appropriate interest. This proposal was accepted 'so far as the Managers are at liberty to accept the said offer'. Clearly legal advice had caused them to rescind these resolutions at their next meeting on July 4th, and Mr Fletcher, the solicitor, was instructed to press Gray's solicitors for a speedy conclusion. In spite of his efforts it was not until November 1809 that the executors of Gray's estate were able to inform the Receiver that funds were available from sales of property for the debt to the Royal Institution to be settled before the Master of the Court of Chancery in the presence of solicitors from both parties. The Court approved the Institution's claim against Gray's estate but payments were further delayed because there was an outstanding mortgage on part of Gray's property and the mortgagee, who had a prior claim, refused to sign any leases until his claim had been settled. That settlement was reported to the Managers in March 1810 but it was not until August of that year that £1517, the amount due up to Michaelmas 1809, was paid over. The balance of interest due up to the date of payment, and the 1809 renewal fine of £370, were both to be paid later by the Receiver. This was not by any means the end of the Institution's problems with their tenants at this period. Not only were payments of the 1809 lease renewal fines from both Nos. 22 and 23 still outstanding in January 1811 but the Managers had had to threaten distress proceedings in respect of unpaid ground rents.[15] The debts owing on No. 23 were evidently settled since the lease to the executors of the Gray Estate had been renewed before the end of 1813.[16] Thus, at the cost of losing an opportunity of extinguishing within 20 years the perpetually renewable underlease on 23 Albemarle Street, the Royal Institution had resolved its immediate financial problem. Subsequent holders of that underlease never again failed to pay the lease renewal fines though, more than once, they had to be reminded that they were in arrears. As a result the Institution, as will later be seen, was compelled by a change in the law of property in1924 and 1925 to grant an underlease on that property of 2000 years from Michaelmas 1921 at a fixed ground rent which, in present terms, is derisory. A great opportunity was lost for what, at that time, were good and sufficient reasons.

In the case of the lease on 22 Albemarle Street a similar opportunity to escape from a perpetually renewable lease offered itself and was fortunately taken. When the Institution bought the head lease from the Mellish estate Mr Swinton, the underlessee, had not then paid his 1795 lease renewal fine of £240 which then became a debt to the Institution. Evidently this had subsequently been paid but the fine due in 1809 was still outstanding early in 1812.[17] The solicitor to the Institution was asked to make further enquiries but it was not until the following year that the Managers were informed that Mr Swinton had died and that his executors wished to renew the lease in favour of Spiller's nephew, Mr James Spooner, who had inherited his uncle's interest in No. 22,[18] by paying off the 1809 fine with interest. This proposal was accepted and a new underlease to James Spooner was sealed on 21 June 1813 to run for 61 years less one quarter from Michaelmas 1809 on the same perpetually renewable terms as the earlier lease to Swinton. The first renewal date on this new lease was 1823.

When that date arrived the Royal Institution was due to pay its own lease renewal fine of £600 at Michaelmas. It was again in financial difficulty over this and other expenses, and, to meet these, the Managers considered[19] the sale of stock purchased with an earlier donation from John Fuller. Before this step was taken they were informed[20] that loans of up to £4000 would be made available without the need to break into that Fullerian fund. Daniel Moore, the newly appointed solicitor to the Institution, advanced the £600 required to pay the lease renewal fine and repayment of this short-term loan was authorized at the meeting on the following December 1st. At that meeting the Managers heard that James Spooner was unable to pay his renewal fine of £240 on the underlease for No. 22 although 'he knew that it was in his interest to have the lease renewed but at present the rent of that house is all he has to maintain his large family and he has not got the means to pay'. In spite of an appeal from Spooner through his solicitor to renew the lease subject to payment of the fine as soon as he was able, the Managers decided[21] that Spooner had forfeited the right to renewal and that the lease on No. 22 would revert to the Institution in 1870.

The prospect of this reversion was of great interest to the Royal Institution and, in 1862, William Pole, the Treasurer at the time, presented a report, accompanied by a plan (Fig. 8), to the Managers concerning matters which required attention if the full value of the reversion was to be secured. The main part of No. 22 lay within land to which the Institution held the headlease, but a bay built at the rear of the house encroached upon land which had been part of the land leased from the City by the Duke of Grafton in 1767. By 1851 this perpetually renewable headlease was held by a Mr Robert Elliott and partners. The Paynter family had, much earlier, obtained an underlease from the headlessee, on similar terms, for 165–167 New Bond Street and 16 Grafton Street, a part of which stands upon the strip of land which had been leased by Edward Gray to Mr William Paynter (see Fig. 8).

By 1818 a Mr Francis Baldry, mercer, was living at 166 New Bond Street,[22] evidently with a renewable underlease on the house and land at the rear from the Paynters. In order to protect his own position James Spooner had negotiated a term lease from Baldry in 1828 for land at the rear of 22 Albemarle Street to legitimize the existence of his bay window and to provide himself with a back yard. Pole pointed out that this lease was due to expire at Midsummer 1870 and that there was no automatic possibility of renewal. Furthermore, Baldry had died and his executors were expected to sell his underlease on No. 166 within the next few years upon the death of his elderly widow. It was essential that the Royal Institution should then acquire it. If not, any new owner would be entitled to demand that the rear bay to 22 Albemarle Street should be demolished and to develop the yard at the rear irrespective of the wishes of the Institution.

There was a further complication in that the Alfred Club, which had held the underlease on 23 Albemarle Street from Gray's successors in title from 1808, had rented No. 22 from James Spooner, together with the land parcel at the rear described in the 1828 lease from Baldry, to form an extension to the Club at some time before 1841.[23] That Club had ceased to trade in 1862 and its assets were held by trustees who were paying a rent to Spooner of £284 p.a. in respect of the house and the additional land. To reduce their cash outflow they had sublet that property to Charles Asprey, then trading at 166 New Bond Street, for £260 p.a. Asprey had already built a 'glass room' at the rear of No. 22 and was warned by the Managers in March 1863 that, at the expiry of the Spooner leases in 1870, he would be expected to return the premises to their original form as a dwelling house. However, Asprey then made it clear that he was anxious to obtain an extension of his lease both on No. 22 and on the land at the rear but that he did not wish to compete with the Institution in the bidding when the Baldry lease came up for sale. Naturally the Institution shared that view and set up a sub-committee under Pole's chairmanship with two objectives, namely, to preserve the rights of the Institution to the land leased by Spooner from Baldry, and to avoid any possibility of being forced to offer a new perpetually renewable lease after 1870 on that land and upon No. 22 itself. The sub-committee then recommended that the Institution should offer a new lease on No. 22, to run for 21 years from 1870 at a rent of £200 p.a., to Charles Asprey, subject to a number of conditions, and that the Baldry lease had to be acquired either by the Institution or by Asprey. If Asprey was to be the successful bidder then the land leased by Spooner at the rear of No. 22 should subsequently be offered for sale to the Institution. If, on the other hand, the Institution obtained that lease, thereby acquiring an interest in 166 New Bond Street, that interest would either be sold to him or, if he preferred, an underlease would be offered to him. These agreements were to bind both parties, their executors, administrators, and assigns without time limit.

The sale of the Baldry lease by auction was announced on 1 April 1865 as one lot. It was a perpetually renewable underlease, originally from Mr William Paynter, at a fixed ground rent of £39 p.a. and a lease renewal fine of £195, the next renewal date being Midsummer 1865. Asprey had been involved in the deliberations of the sub-committee at their meetings on 22 and 28 March 1864. At the first of these he had offered, if the Institution were to buy the lease, to rent the whole property (166 New Bond Street and part of 22 Albemarle Street) at 4% of the purchase price and to pay both the ground rent (£39 p.a.) and £14 p.a.

towards the lease renewal fine. At their meeting on March 22nd the sub-committee authorized Asprey to attend the auction and to bid up to £7000, a figure revised at the second meeting on Asprey's advice to £9000.[24] It was agreed that he should be free to choose whether to buy for himself or on behalf of the Institution, since the 1863 agreements preserved the interests of both parties irrespective of the purchaser of the lease. Asprey secured the lease for £7900, paying a 10% deposit. He was not prepared to pay the balance himself at that time, asking for the deposit to be repaid to him, thus leaving the Institution to be the purchaser. However, he wanted to retain the option to purchase 166 New Bond Street itself for longer than had been allowed for in the 1863 agreements and, in addition, to have a guaranteed right to the renewal of his leases. The Managers agreed to extend the option to June 1865 but refused any guarantee of lease renewal. This substantial purchase was financed by the sale of £8800 of Consols at 90. These had given an income of £264 p.a. as compared with the 4% of the purchase price for the Baldry lease (£316 p.a.) which Asprey had agreed to pay for the a lease of 27 years less one quarter from Christmas 1864.[25] From 1870 this would be supplemented by the rental from Asprey of £200 p.a. for No. 22 rather than the £48 p.a. paid by Spooner during his lease. After 1870, therefore, the investment income of the Institution increased by £204 p.a., a substantial sum for the period, and a perpetually renewable underlease from the Paynter family had been secured. The negotiations which led to this successful conclusion were due in large part to the legal skills of William Pole and his achievement profits the Royal Institution at the present time and will continue to do so in the future.

In May 1894 the leases which had been held by Robert Elliott and his partners were to be auctioned and Lot 4 comprised the City headlease on the major part of 16 Grafton Street and 165–167 New Bond Street. Thomas Lupton, who was then the solicitor acting for the Royal Institution, recommended that efforts should be made to purchase this lease in order to strengthen the position of the Institution in relation to the underleases already held on 166 New Bond Street and part of 22 Albemarle Street.[26] This was agreed without much discussion and Lupton was authorized to bid up to £3000, a figure thought to be above the likely market value. He was also asked to let the Paynter family know of the intentions of the Institution in the hope that they could be persuaded not to bid. Although he was successful in this, a third potential bidder forced him to go to 3000 guineas (£3150) in order to get these head leases. The ground rent due to the City was £52 p.a. with a lease renewal fine of £225 and the prime underlease to the Paynters carried a ground rent of £119 p.a. and a lease renewal fine of £595. The net yield to the Institution was therefore about 3% p.a., comparable with that on the stocks which had been sold to finance this lease purchase which was finally completed in 1895.

After that date the Institution held perpetually renewable City leases on the east side of Albemarle Street northwards from No. 20 up to Grafton Street, the whole of the south side of Grafton Street between Albemarle and New Bond Streets, and 165–167 New Bond Street. The only leasehold property used by the Institution was 21 Albemarle Street, and, with the exception of 22 Albemarle Street and 166 New Bond Street, the rest of the property was held by tenants on perpetually renewable leases from the Royal Institution on fixed terms. Any changes to buildings on the City Conduit Meads estate required a licence to be granted by the freeholder to the head leaseholder, and, as early as the 1840s the City seems to have found this an administrative burden. For example, in 1841, the Alfred Club, which had occupied 23 Albemarle Street since 1808, had taken over No. 22 and wished to open up communication between the two adjacent buildings. When the Royal Institution applied on the Club's behalf for the licence it was granted with the comment that the City authorities had little interest in the Conduit Meads estate because of the renewable leases and did not wish to interfere. This lack of interest was shown both in 1889 and in 1898 when the City attempted to persuade head leaseholders to avoid the administrative problems associated with the lease renewal process by amortizing the fines, either by payment of a capital sum or through an increase in the ground rent. The Institution did not take up these offers; instead, the Managers arranged for the lease renewal dates and fines to be painted on the walls of the Managers' room as an *aide-mémoire*.

Perpetually renewable leases became illegal after two Law of Property acts in 1924 and 1925, and 1921 was therefore the last date on which lease renewal fines on the Conduit Meads properties had to be paid. The City then proposed to convert all such leases into term leases of 2000 years from 29 September 1921. The financial terms first proposed for this change by the City were strenuously opposed by many of the leaseholders, including the Royal Institution. Eventually the matter went to arbitration and an agreed solution was not reached until early in 1928.[27] Where headleases had granted perpetually renewable underleases these also became 2000-year leases with fixed rents based on eighteenth-century values uprated to take account of the disappearance of the periodic lease renewable fines. As a result, with the exception of 22 Albemarle Street and 166 New Bond Street, acquired by purchase of the Baldry lease in 1865, the underleases granted by the Royal Institution yield little by way of rental income and will not be surrendered until that distant time (AD3921) when all the Conduit Mead leases revert to the City of London. The Institution does retain some formal control in the sense that, in principle, any fabric alterations to the buildings on the land, and any changes in title require licences from the Institution and from the City authorities.

The successors to Charles Asprey have occupied 22 Albemarle Street and 166 New Bond Street, with the land behind No. 22 since 1870 on a series of fixed-term leases the last of which were negotiated in 1984. For many years they had sublet the upper floors of No. 22 to a series of tenants which, in the nineteenth century, included the Royal Asiatic Society, the London Mathematical Society, and the British Association,[2] though when Asprey's wished to renew their lease in 1924 the Institution contemplated taking over those upper floors for library use.[28] That provision does not seem to have been inserted in those 1924 leases and extra space for the library was not obtained until the purchase of the freehold of 19 Albemarle Street in 1935. These Asprey leases do provide a substantial income to the Institution which is open to review at defined intervals. This fortunate outcome arises from the failure of James Spooner to pay his lease renewal fine on No. 22 in 1823 and the probable action of an eighteenth-century builder in ignoring the details of lease boundaries when the house was built.

The freehold properties

20 Albemarle Street was the first of the three to be acquired. It came as a gift from Ludwig Mond, who had made his fortune from the chemical industry, in order to improve the facilities for research in the Institution. That gift, together with a substantial endowment, made possible the establishment of the Davy Faraday Laboratory by the terms of a Trust Deed. Mond's generous proposal was made to the Managers in 1894,[29] and the conveyance of the building, together with the Trust Deed, was sealed by the Institution and by Mond in the summer of 1896.

By that act 20 and 21 Albemarle Street became a single establishment for the first time since John Roberts bought Grantham's freehold and leasehold property in 1755. No. 20 had remained in private occupation even after Roberts's tenant William Mellish had moved into No. 21 in 1775 until it was sold in 1834 to Mr John G. Chaplin for use as an extension of the Clarendon Hotel by the executors of the Revd Sir John Robinson, Bt who had lived there since 1806.[30] Since the early years of the nineteenth century that hotel had occupied the house built around 1703 for the Duke of Grafton and the leasehold building to the north, both of which fronted on to New Bond Street, see Figs 6 and 7. The whole complex is numbered 169 in the mid-nineteenth-century street directories. As can be seen from the Figures there was a garden between the leasehold extension and the boundary of the Royal Institution property. The leaves from the trees in that garden as they settled in the autumn on the skylight of Michael Faraday's basement laboratory hindered his researches so much by reducing the level of natural light that he was led to complain to the Committee of Managers. This was a problem which the Committee did not attempt to solve, merely suggesting that Faraday should approach Mr Chaplin! A more important point arises from the minutes of meetings of the Managers between February and July 1834 since Chaplin requested the transfer of a perpetually renewable lease on 'a small piece of land and three vaults or cellars partly under No. 21'. There was a legal necessity to do this but at subsequent lease renewal dates the Managers occasionally expressed irritation about their lack of control over that part of the property of the Institution. Mond's gift of the freehold of No. 20 included the return of that perpetual lease to the Institution, thus solving a long-standing difficulty.

The Clarendon Hotel continued in business until 1870 and, for a period spanning the middle of the century, had a great reputation for its French cuisine apart from being a residential hotel with a distinguished clientele. After its closure No. 20 seems to have reverted for a time to its status as a private residence but later housed the New Oxford and Cambridge Club for a few years immediately before Mond bought it.

After the gift, Nos. 20 and 21 were reconnected at the first-, second- and third-floor levels. Mond paid for the two connecting doors to be made at the end of the Main Library leading into the Writing Room and the Long Library. Further costs associated with the carpeting, furnishing, and the fittings of these rooms were borne by the Institution.[31] The Resident Professor, Sir James Dewar, was allocated two rooms on the second floor of No. 20, rooms which have since formed part of the Director's apartments, partly because of an anxiety on the part of the Managers about the dangers of leaving No. 20 unoccupied at night. The work of making the connection and the associated 'structural alterations and repairs necessitated thereby' cost about £700, borne by the Institution, and Dewar contributed a further £250 for the ensuing redecoration.[31] This allocation carried an unexpected consequence. In April 1897 Dewar reported that he had received a quarterly rate demand of £84 as the sole resident of No. 20. Neither he nor the Managers had expected this, and, as a consequence he was given an annually renewable lease from Christmas 1896 at a rental of £100 p.a., a device adopted to confine the liability for parochial rates to these rooms. Notwithstanding these covenants it was resolved[32] that 'he, [Dewar] should not pay this rental except for a peppercorn as and when demanded' and the Managers agreed to pay all rates and taxes levied on these rooms. A later lease in similar terms was offered to Sir William Bragg on his appointment but this practice was later dropped.

In May 1935 the Managers learnt that the freehold of No. 19 was available with vacant possession at an asking price of £25 000. They commissioned a professional survey and valuation which revealed that the property was in indifferent condition, the interior being in a poor state of repair, and that the freehold was worth little more than the site value which was estimated at £22 500.[33] The Managers inspected the house and authorized the Treasurer to negotiate the purchase. On June 3rd he signed the agreement to purchase at £22 000, the lowest figure which the vendors would accept. This sum was financed by a loan from Drummonds Bank at favourable terms,[34] and, by November, it was decided to use the third and fourth floors only for Institution purposes and to let the rest at commercial rates as soon as the necessary alterations and repairs could be carried out. These included repairs to the roof, the insertion of a shopfront at ground-floor level, the provision of modern lavatory accommodation, and internal redecoration. In addition an entranceway into No. 19 from No. 20 and a service lift had to be constructed. This work was finished in March 1936 and the basement, ground, first, and second floors were fully let by July on 14-year leases with options to terminate at 7 years. The initial total rental income was £1215 per annum with the landlord responsible for the rates. By November the top two floors were in use by the Institution for an extension to the library with areas for the display of models and for storage purposes. In the intervening years there have been changes in the use of the building and up to recently the Davy Faraday Laboratory occupied the first and fourth floors, the rest being let. It has now (1997) been decided to withdraw all Institution activities into Nos. 20 and 21 and to let the whole of No. 19 for income generation.

During the Second World War L. Rome Guthrie, who had been consulting architect to the Institution since 1912 and had been responsible for the extensive rebuilding of No. 21 during the 1930s, produced a scheme for extra floors on Nos. 19 and 20, and that kind of optimism about expansion continued into the 1960s. From 1960 onwards the Managers were even considering the possibility of demolishing 19 and 20 Albemarle Street and of replacing them with a modern building. A Development Committee was set up in 1962,[35] and Russell, Orme and Partners were instructed to prepare plans on this basis. In this optimistic climate of opinion an opportunity arose in 1963 to purchase a short lease on No. 18 from Blanch Littler (Beauty) Ltd, a salon run by a fashionable hairdresser of the period, Mr 'Teazy-Weazy' Raymond. The company had a 14-year lease from Christmas 1955 at a rent of £2850 per annum, and was offering the residue of the lease for around £11 000. At the Managers' Meeting of 31 October 1963 it was suggested that, if the property could be sublet for £6000 p.a., purchase of the short lease on such terms would be an economic proposition. At the end of the period the purchase of the freehold might be possible. If not, the Court would have power to grant a lease extension of up to 14 years. The capital required could be obtained by realizing investments, though disposal of 'the Asprey freehold' (presumably the 2000-year leases on 22 Albemarle Street and 166 New Bond Street) had briefly been considered. The Managers accepted this investment analysis and authorized purchase for a sum not exceeding £11 000. Initially there was some delay in implementing this decision because of uncertainty about the freehold,[36] but the licence to assign the remainder of the lease was sealed at a special meeting of the Managers on December 20th. The purchase price was agreed at £10 750, a sum financed by the sale of gilt-edged stock. In February 1964 it was reported that the John Akerman photographic firm had rented the whole house at £6000 p.a. for the duration of the lease. On that basis, the investment analysis showed that the expected rental income would exceed the total of the purchase price, the loss of investment income, and other foreseen costs by over £3000 over the six years remaining.

It soon became apparent that there were problems. The lease to Blanche Littler from the freeholder (Col. J. R. H. Orlebar) gave consent only for the building to be used as a hairdressing salon. Any change of use required the freeholder's consent, and Orlebar had failed to respond to requests for permission, both for the sublease from the Institution to Akerman and for the subsequent subleasing by Akerman of the ground and first floors to Cornelius Spink for use as an art gallery.[37] Eventually these permissions were obtained and in July 1965 Akerman was given a further licence to sublet part of the third floor. By June 1966 Akerman was in arrears with his rent and it was resolved to take immediate steps to gain possession.[38] Later that year Akerman Ltd went into voluntary liquidation, owing £550 in rent to the Institution, and there was little prospect of recovering more than 10% of this.[39] By October the Institution had regained possession and had taken over Akerman's sub-tenants.

Meanwhile enquiries about purchasing the freehold were continuing and on 4 May 1964 the Managers were informed that an offer of £75 000 was likely to be accepted. That offer was made but no reply was received. The matter lay in abeyance until early 1967 after the Secretary had made a special journey to Orlebar's home in Bedfordshire. The Colonel was away but his wife entertained the Secretary to tea and promised to remind her husband of the Institution's interest in the freehold when he returned. The visit bore fruit and in March the Colonel expressed an interest in a firm offer provided that the purchase could be completed by 5 April 1967. A valuation persuaded the Managers that £85 000 should be offered initially with the possibility of negotiating a price not exceeding £92 000. The freeholder drove a hard bargain and, in the end, £92 000 was required to seal the purchase. In total, the acquisition of No. 18 cost the Institution over £104 000—a major investment.

It seems that no detailed structural survey had been carried out prior to the purchase, and by the end of 1967 it was apparent that major repairs were needed at second-floor level before letting would be possible. Matters moved slowly and it was not until October of the following year that a refurbishment contract of £8279 had been let with the expectation[40] that the house would be ready for tenants in the second quarter of 1969. During the contract it was found that the second and third floors were only one third of the strength required for office use and that remedial work would cost at least £1329.[41] In spite of this the repairs were completed by May 1969, almost three years after Akerman had gone into liquidation with the sudden loss of a major slice of the rental expected in 1964. The original investment analysis was, with the benefit of hindsight, highly optimistic.

No. 18 has never been used for Royal Institution purposes and an early suggestion[42] that the development scheme then under consideration for 19 and 20 Albemarle Street from Russell, Orme and Partners should be extended to include No. 18 seems to have been rejected at an early stage. In the event, the whole of this development scheme was abandoned in 1966[43] when £6225 was authorized for payment to Russell, Orme and Partners for the schemes and drawings they had prepared, and it was resolved that no further fees should incurred on that project without specific approval. By that time, refurbishment of the main Lecture Theatre and the construction of the Bernard Sunley theatre had been already agreed and these projects were completed by the end of the decade. In subsequent years 18–21 Albemarle Street have all become listed buildings and there can no longer be a question of demolition and subsequent redevelopment, quite apart from the obvious problem of finance.

The choice of No. 21, rather than possible alternatives, as the home for the Royal Institution must have been influenced by the low outgoings on the property consequent upon William Mellish's policy of leasing most of the Grantham garden as two building plots prior to 1799. If it had not been for an apparent surveying error leading to the encroachment of No. 22 on to land not included in the Royal Institution lease from the City of London and hence in the underlease from Mellish to Swinton, there would have been no incentive to purchase the Baldry lease in 1864 or to acquire the headlease on 'the greater part of 16 Grafton Street, Nos. 165–167 New Bond Street, and a part of No. 22 Albemarle Street' to borrow words from the relevant lease document. It is unlikely that the freehold of 20 Albemarle Street could ever have been acquired without Mond's generosity and, without that freehold, the mere availability for subsequent purchases of the freeholds of Nos. 18 and 19 would have had little interest for the Institution. One could, if asked why the property portfolio had been acquired, echo the words used by the late Harold Macmillan in explaining the influences upon his policies as Prime Minister, 'events, dear boy, events!'. Our founders might well have been astonished at the outcome.

Acknowledgements

Much of the information on the early history of the site has been derived from the detailed research upon which B. H. Johnson based his excellent book,[1] supplemented by other information in the City of London Guildhall Library, particularly on the details of the Normanby lease and the early developments at the north end of Dover and Albemarle Streets. I am indebted to the City of London for permission to copy plans in the Guildhall Library on which Figs 3 to 7 inclusive are based. Johnson's publishers, John Murray, have been unable to trace the copyright owners of Reference 1 but I would be pleased to be put in contact with them.

References

1. Based upon B.H. Johnson, *From Berkeley Square to Bond Street*, John Murray, London, 1952
2. Quoted in Edward Walford, *Old and New London, Volume 4*, Cassel and Co., London, 1887
3. A.D.R. Caroe, *The House of the Royal Institution*, The Royal Institution of Great Britain, 1963
4. Minutes of Meetings, Committee of Managers, 6 April 1799
5. *Ibid*, 21 March 1800
6. *Ibid*, 1 March 1802
7. *Ibid*, 7 December 1801
8. *Ibid*, 11 May 1801
9. *Ibid*, 5 May 1800
10. *Ibid*, 9 May 1811
11. *Boyle's Court Guides*, 1796; 1802
12. G.E.C., *The Complete Peerage, or a History of the House of Lords and all its Members from the earliest times.* Revised and enlarged edn, St Catherine's Press, London, 1929
13. Minutes of Meetings, Committee of Mangers, 23 November, 28 December 1807
14. *Ibid*, 23 January 1808
15. *Ibid*, 14 January 1811
16. *Ibid*, 13 December 1813
17. *Ibid*, 16 March 1812
18. *Ibid*, 1 February 1813
19. *Ibid*, 12 May 1823
20. *Ibid*, 7 July 1823
21. *Ibid*, 1 December 1823
22. *Westminster Poll Book*, 1818
23. Minutes of Meetings, Committee of Managers, 1 February, 1 November 1841
24. *Ibid*, 18 April 1864
25. Licence in Mortmain, 1894
26. Minutes of Meetings, Committee of Managers, 22 May 1894
27. Letter to Thomas Young from Mr Mander, Solicitor to the Royal Institution, March 1928
28. Minutes of Meetings, Committee of Managers, 1 December 1924
29. *Ibid*, 2 July 1894
30. *Boyle's Court Guide*, 1806
31. Minutes of Meetings, Committee of Managers, 7 December 1896

32. *Ibid*, 5 April 1897
33. *Ibid*, 20 May 1935
34. *Ibid*, 1 July 1935
35. *Ibid*, 7 September 1963
36. *Ibid*, 2 December 1963
37. *Ibid*, 11 May 1964
38. *Ibid*, 4 July 1966
39. *Ibid*, 3 October 1966
40. *Ibid*, 4 November 1968
41. *Ibid*, 2 December 1968
42. *Ibid*, 2 December 1963
43. *Ibid*, 7 February 1966

H. J. V. TYRRELL

Born in 1920, he studied chemistry at Jesus College, Oxford and, on graduation in 1942, he worked in the chemical industry until appointed as an assistant lecturer in the Department of Chemistry at Sheffield University in 1947. There he developed research interests primarily in transport processes in fluids and saw the department grow in size and reputation before moving in 1965 to a chair in chemistry in what was then Chelsea College of Advanced Technology but soon to become a constituent college of the University of London. Later, he served the College as Vice-Principal and finally as Principal up to the time when it merged with Queen Elizabeth College and King's College in 1985, becoming Vice-Principal of the combined College until retirement in 1987. He became a Member of the Royal Institution in 1967 and served on the Committee of Managers before serving as Honorary Secretary from 1978 to 1984, a period marked by major changes in the constitution and bye-laws of the Institution. He later became a member of the newly formed Council, of the Finance Committee, and for some years was Chairman of the Buildings Working Party, a responsibility which stimulated the research leading to this work.